Interpretation of Landforms from Topographic Maps and Air Photographs Laboratory Manual

Interpretation of Landforms from Topographic Maps and Air Photographs Laboratory Manual

DON J. EASTERBROOK
Western Washington University
Bellingham, Washington

DORI J. KOVANEN
University of British Columbia
Vancouver, British Columbia, Canada

Prentice Hall, Upper Saddle River, New Jersey 07458

EXECUTIVE EDITOR: *Robert McConnin*
PRODUCTION EDITOR: *Alison Aquino*
MANUFACTURING MANAGER: *Trudy Pisciotti*
COVER DESIGNER: *Kiwi Design*
ART DIRECTOR: *Jayne Conte*

Printed in the United States of America
10 9 8 7 6 5 4

ISBN 0-13-976002-4

Prentice-Hall International (UK) Limited, *London*
Prentice-Hall of Australia Pty. Limited, *Sydney*
Prentice-Hall Canada, Inc., *London*
Prentice-Hall Hispanoamericana, S.A., *Mexico*
Prentice-Hall of India Private Limited, *New Delhi*
Prentice-Hall of Japan, Inc., *Tokyo*
Pearson Education Asia, *Singapore*
Editora Prentice-Hall do Brasil, Ltda., *Rio de Janeiro*

Contents

The omission of Ch. 8 is intentional. The laboratory exercises are numbered to correspond to the chapters in Surface Processes and Landforms, *and Chapter 8 (Tectonics) has no exercises.*

Preface

The exercises in this laboratory manual have evolved over a period of many years of use in the laboratory section of the geomorphology course at Western Washington University. The exercises assume that the user has a basic understanding of topographic maps, aerial photographs, map symbols, contour lines, topographic profiles, and geologic cross-sections. These topics are presented in all introductory physical geology textbooks.

The intent of the exercises is to develop the ability to interpret the landforms on any map or air photo, utilizing the method of multiple working hypotheses. Therefore, the questions are posed not so much as an exercise in naming or describing landforms, but rather to foster a mental process in problem solving.

Because of size and space restrictions, only portions of the 7.5- or 15-minute quadrangles may be presented at appropriate scales. For those preferring to work with the original, full-size maps, Appendix A lists the map name, location, scale, and contour interval for each exercise. The contour interval on the topographic maps are in feet and therefore the metric system is not used in this manual.

Interpretation of Landforms from Topographic Maps and Air Photographs Laboratory Manual

Introduction

Basic Concepts

The study of geomorphology is based on a number of fundamental principles, some of which are shared with other scientific disciplines and some which are unique to geomorphology. Because the earth's topographic features are a mixture of landforms being formed at the present time and others that have been shaped in the past by processes no longer active, geomorphology embraces the investigation of both the mechanics of modern processes and the historic influence of geologic time. The former includes an understanding of the physics and chemistry of surface processes that generate landforms, and the latter adds the element of time to landforms that evolve over periods of time too long to study in the context of modern processes. The origin of landforms can be related to a particular geologic process, or set of processes, and the landforms thus developed evolve with time through a sequence of forms having distinct characteristics at successive stages (Davis, 1909).

The landforms that we observe today have evolved over a period of time as a result of surface and subsurface processes. The factors that govern the development of landforms in any area are:

1. **Geomorphic process**; each process bestows distinctive features on the landscape and develops characteristic assemblages of landforms from which the origin of the forms can be identified.

2. **Stage of evolution of landforms**; landforms evolve with time through a continuous sequence of forms having typical features at successive stages of development, largely as a result of continuous changes in processes and rates as time goes on.

3. **Geologic structure**; some landforms are a result of tectonic disturbance of the earth's crust. Such landforms are produced either by directly offsetting the land surface or by secondary erosion of rocks of differing resistance.

Suggestions For Interpreting The Origins Of Landforms

Interpretation of landforms involves understanding fundamental concepts of surface processes and the application of basic logic. Some suggestions to aid in interpretation of the maps and air photos in this manual include:

1. Learn the basic approaches for determining types of topographic expression.

2. Learn the geologic processes by which the landforms are shaped. These processes include the work of the major geologic mediums such as running water, groundwater, volcanism, glaciers, waves, wind, weathering, and mass wasting. Each of these processes leaves its mark on the landscape in the form of one or more characteristic landforms.

3. Learn to recognize the stages of development of landforms and how they evolve with time.

4. Familiarize yourself with criteria for recognition of topographic expressions of geologic structures.

5. Utilize the method of *multiple working hypotheses* in interpreting landforms. Get in the habit of systematically going through the following steps:

a. **Observation of facts**. Note the shape, size, orientation, and composition of landforms and their association with other landforms.

b. **Develop multiple working hypotheses** for the origin of the observed landforms. Several alternative interpretations may be possible to explain the observed data. Once a list of all possibilities has been compiled, attention can be focused on critical evidence needed to reject or prove specific hypotheses of origin.

c. **Test the hypotheses.** (i) Reject all hypotheses that do not fit the observed facts, (ii) develop a list of possible solutions to the problem at hand, (iii) seek out specific evidence to support or reject any of the remaining hypotheses, and (iv) test each hypothesis against all of the evidence at hand.

An example of applying the method of multiple working hypotheses to geomorphic problems is given in Chapter 2 of *Surface Processes and Landforms*.

Topographic Maps and Aerial Photographs

The origin and evolution of landforms may be interpreted on the basis of the size, shape, orientation, composition, and distribution of topographic features, which may be analyzed from topographic maps and aerial photographs, as well as by direct observation of the land surface. Topographic maps can portray three-dimensional forms on two-dimensional paper and provide accurate data concerning size, shape, orientation, distribution and elevation of the landforms. The relationships of a landform to nearby associated landforms may readily be seen on topographic maps.

Topographic maps are of such great importance in geomorphic interpretation that students need to thoroughly understand them. Most of the exercises in this manual assume basic map reading skills. A brief review is presented in the appendix.

Topographic Map Scale

When analyzing landforms from topographic maps, careful attention must be paid to the map scale because interpretation of surface features involves consideration of size, as well as the other factors previously mentioned. Map scale is the relationship between distance on a map and the corresponding distance on the ground and may be expressed as a numerical ratio or shown graphically by bar scales. A map which is drawn at a scale of 1:125,000 indicates that one inch on the map equals 125,000 inches of ground distance, or any other unit of measured map distance equals 125,000 of the same units of ground distance.

Maps are made in a great variety of scales, depending on how much of the earth's surface is to be shown on a given size of paper and on how much detail is desired. U.S. Geological Survey maps are generally available at following scales: 1:24,000 (7.5-minute quadrangles), 1:62,500 (15-minutes quadrangles), 1:250,000, 1:500,000, 1:100,000, 1:1,000,000. For interpretation of most landforms, 1:24,000 maps are preferable, but regional relationships show up well on 1:62,500 and 1:100,000 maps.

To measure distance from maps, the map distance between two points is measured and then converted to ground distance. For example, consider two points that are 5 inches apart on a map drawn at a scale of 1:24,000. Since 1 inch on the map equals 24,000 inches of ground distance, 5 inches equals 5 x 24,000 inches of ground distance. To put this large number into more usable terms, it may be divided by 12 to convert it into feet, and then further divided by

5,280 to convert it into miles. A map distance of 5 inches at this scale thus represents a true distance of slightly less than 2 miles.

The apparent size of a given feature on a map of differing scale is quite different, and the amount of detail shown may be significantly different. Much greater detail is shown on a 1:24,000 map than on a 1:62,500 or 1:250,000 map. Although details of landforms become indistinct on maps of smaller scale, such maps are still useful because they show relationships of landforms to one another over a larger area. Thus, in analyzing landforms in an area, the choice of map scale will depend largely on how much detail is necessary for the purpose of the investigation.

Converting From One Scale to Another

Conversion of one scale to another to is often necessary to estimate the size of a landform or to compare features on maps or photos having different scales. The following are examples of some conversion problems.

1. Convert a map scale of 1:125,000 to a scale using inches per mile.

    1 inch on the map = 125,000 inches on the ground
    1 mile equals 5,280 feet, and 1 foot equals 12 inches, so
    5,280 x 12 = 63,360 inches in one mile
    If we divide 125,000 inches by the number of inches in a mile, we arrive at:

    $$125,000" \times \frac{1 \text{ ft.}}{12 \text{ in.}} \times \frac{1 \text{ mi.}}{5280 \text{ ft.}} = 1.97 \text{ mi.}$$

    or, for practical purposes, 1 inch = 2 miles

2. If a landform that is 3" long on an air photo is 2 miles long on a map, what is the photo scale?

    $$\frac{1}{X} = \frac{3"}{1} \times \frac{1 \text{ mi.}}{5280 \text{ ft. (2)}} \times \frac{1 \text{ ft.}}{12 \text{ in.}} = \frac{1}{42,240}$$

Air Photographs

A wide variety of aerial photographs from aircraft and satellites have become available for use in interpretation of landforms. In addition to black-and-white and standard color photos, infrared, side-looking radar, x-ray, and various types of color photos may be used to identify surficial features that are not obvious from the ground.

Air Photograph Scale

The scale of an aerial photograph is determined by the altitude of the plane and the focal length of the camera—the higher the altitude of the plane, the smaller the scale. In the diagram on the following page, the focal length of the camera is f (p$_1$ L) and the altitude is H (LP).

Because a1, p1, L and APL in figure on the following page are similar triangles, image distance a1-p1 is proportional to ground distance AP, and p1 L is proportional to PL; that is, focal length (*f*) is proportional to altitude (H).

For example, if the focal length of a camera is 6 inches (0.5 ft.), and the altitude of the plane is 20,000 feet, the scale of the photos will be:

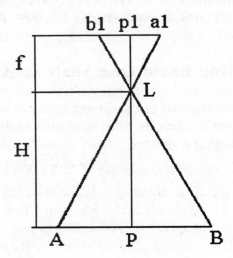

$$\frac{f}{H} = \frac{0.5}{20,000} = \frac{1}{40,000}$$

The scale of aerial photographs determines the amount of detail shown on the photos. On photographs taken from a high altitude, minor features may be too small to recognize, whereas on photographs taken from a lower altitude, regional relationships of landforms to one another may go unnoticed. For reconnaissance work, a scale of 1:60,000 may be satisfactory, but for detailed work, a scale of about 1:5,000 may be more appropriate.

Oftentimes the altitude of the plane and camera focal length are unknown. However, the scale of a photograph may be easily determined by measuring the distance between two prominent points on the photograph and comparing that distance with the true ground distance measured from a topographic map. The ratio of photograph distance to true round distance is the scale of the photo. For example, if two features are 3 inches apart on an air photograph, and if the distance between them is determined from a topographic map to be 2 miles, then the scale of the photograph may be calculated as follows:

3 inches (photograph distance) = 2 miles (ground distance), so

$$\frac{1 \text{ inch (photograph distance)}}{X \text{ miles (ground distance)}} = \frac{3 \text{ inches (photograph distance)}}{2 \text{ miles (ground distance)}}$$

and, 1 inch (photo distance) = 2/3 mile = 3,520 feet (ground distance)

or, 1 inch = (12) (3,520) = 42,240 inch, and the scale is 1:42,240

Steroscopic Air Photo Pairs

We see three-dimensional images in our everyday life because our eyes observe objects from two slightly different points in space, a process known as *parallax*. This same phenomena also occurs if two photographs of the same feature are taken from slightly different positions, as in successive photos taken from an aircraft. If two such photos are aligned so that one eye looks at one photograph and the other eye looks at the second photograph, a three-dimensional (stereoscopic) image will be seen. In viewing air photos stereoscopically, the topographic relief of an area is somewhat exaggerated, a phenomenon that is useful in bringing out details of landforms.

In order to view two air photos stereoscopically, each photo must be oriented so that the same feature can be seen in each photo with one eye. Precise procedures may be used to properly orient photos, but for most stereophotographic use, a less exact, but faster, method is more efficient. The objective is to look at some prominent feature in one photo with one eye, find the same feature in the other photo and look at it with the other eye. If the photos are not properly oriented, two overlapping images will be seen, one with each eye. By moving the photographs back and forth and adjusting them until the two images merge into a single, 3-dimensional image, the appropriate orientation may be determined.

TOPOGRAPHIC MAPS - REVIEW

As a basic review of topographic maps, complete the following.

1.   On a separate piece of paper, define the following terms:

          Contour   Contour interval

          Map scale   Relief

          Profile   Vertical exaggeration

          Latitude   Longitude

2. Make a sketch of the location of the SW 1/4 of the NW 1/4 of sec 12, T29N, R2W.

3. What is the scale (stated as a ratio) of a map where 1inch = 1 mile? Show your calculations.

4. On a map drawn to a scale of 1inch = 1 mile, what distance on the map represents 2,000 feet? Show your calculations.

5. What is the scale (stated as a ratio) of a map where 1" = 2,000'? Show your calculations?

6. On a map drawn to a scale of 1:100,000, what distance is represented by 3 inches? Show your calculations.

7. On a map drawn to a scale of 1:100,000, what distance is represented by 3 cm? Show your calculations.

8. A 4"-long ridge on an air photo is 2 miles on a map. What is the photo scale?

TOPOGRAPHIC MAPS - CONTOUR SKETCHING

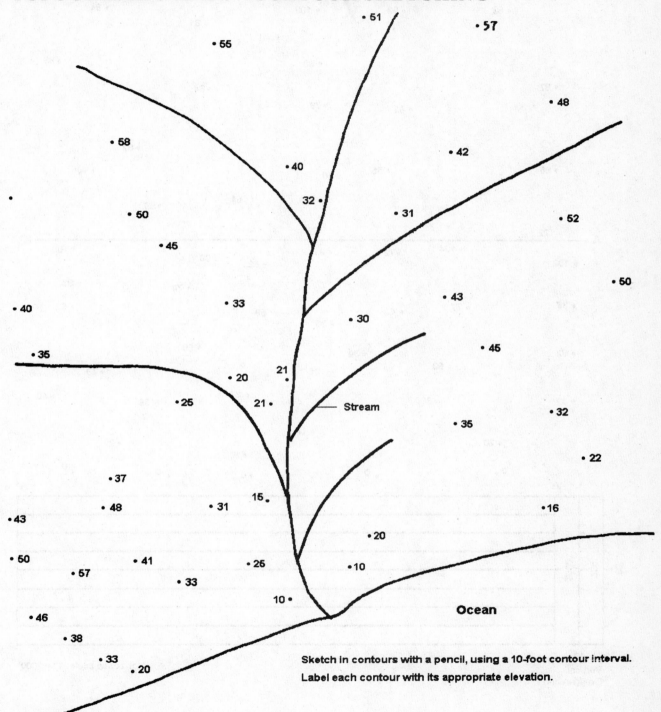

Sketch in contours with a pencil, using a 10-foot contour interval.
Label each contour with its appropriate elevation.

TOPOGRAPHIC MAPS - CONTOUR SKETCHING

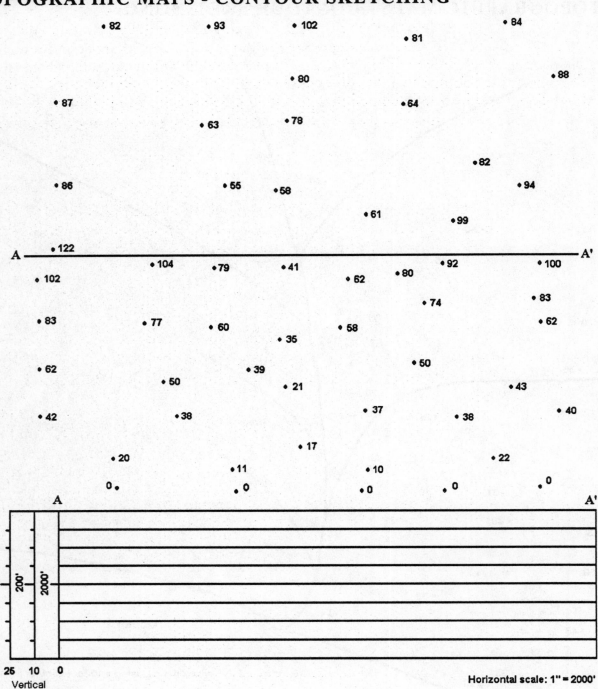

1. Sketch in contours for the elevations shown above, using a contour interval of 20 feet.
2. Using the contours from the map, construct 3 profiles along line A-A'. Profiles with vertical exaggeration of 0 (the same vertical and horizontal scales), 10 and 25.

WEATHERING – SW UTAH

Weathering proceeds at more rapid rates where rock is less resistant or where water may gain access to the rock along joints or bedding planes. The dark rock on the photo is gently dipping shale that overlies light toned sandstone.

1. What makes the well developed linear pattern in the horizontal sandstone? _____
Explain what causes this differential erosion? _____

2. What is the linear pattern not formed in the overlying shale? (Hint - think in terms of relative brittleness of shale versus sandstone.) _____

Rock fin   Accelerated weathering   Arch
at base of fin

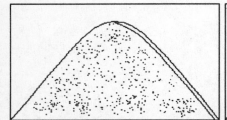  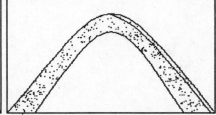

Weathering proceeds along joint planes more rapidly than within the massive sandstone, leaving fin-like ridges of sandstone standing between weathered joint planes, as seen in the photo above.

3. Explain how an arch evolves from joint-bounded rock fins? _____
_____

4. What physical phenomena allow the arch to stand without collapsing? _____
_____

Mass Wasting

Criteria for Recognition of Mass Wasting Features

Mass movement is often the result of oversteepening of a slope, undercutting the toe of a slope, slope that become saturated with water, stress on the material that exceeds its resistance to movement, and seismic shaking. Mass movements may be recognized by the following topographic features:

1. arcuate scarps and hollows
2. springs, ponds, and swamps
3. ground cracks
4. disturbed vegetation
5. offsetting of streams, damming of drainage
6. lobate tongues with hummocky surfaces
7. pressure ridges on the surface
8. benches that tilt into the slope behind them

MASS WASTING – LYDEN, NEW MEXICO

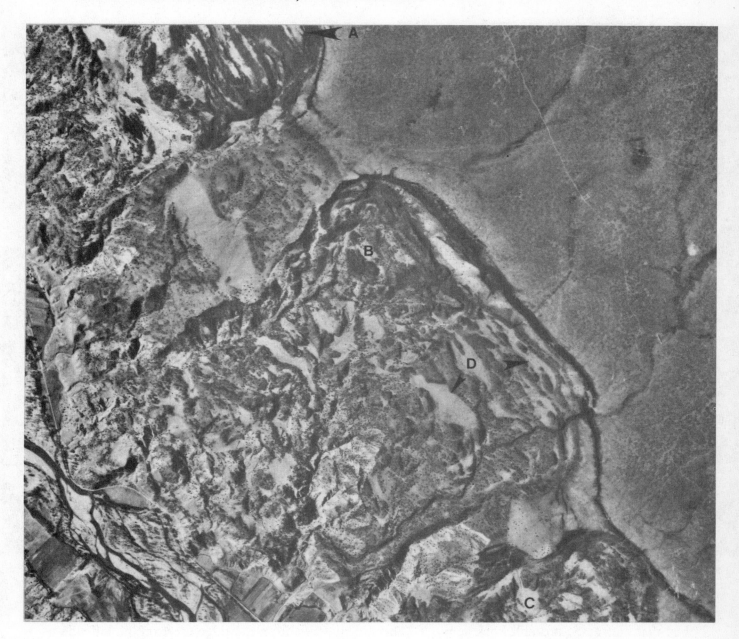

1. The irregular topography at 'A', 'B', and 'C' is caused by slope failures along the steep slopes. What is the dominant type of slope failure? _____

2. Note the short, linear segments of light-toned, flat benches within the irregular topography (D). What causes these benches? _____

_____

MASS WASTING – MADISON SLIDE, MONTANA

The large slope failure shown on this air photo is the Madison slide along the Madison River, Montana. It was triggered by an earthquake in 1959 that fractured a resistant limestone bed that buttressed unstable rock above it. The slide extends from the ridge crest on the left side of the photo to the Madison River near the top of the photo.

1. Draw the boundaries of the slide on the air photo. How wide is it? _____
How long is it? _____

2. What effect has the slide had on the Madison River at its terminus? _____

3. Note the lake at 'A'. If the lake level rises until it tops the slide debris, what may happen?

_____

MASS WASTING – COLORADO

1. What was the direction of movement of the valley-filling landslide at 'A'? _____

2. What effect has it had on the main trunk stream at 'B'? _____

3. What has happened to the drainage in the valley at the distal end of the slide?

_____

How does the slide-generated lake drain? _____

4. Why does it drain across the far side of the slide? _____

5. Judging from the extent of the slide, how mobile was it? _____

Fluvial Processes and Landforms

Topographic Criteria for Recognition of Fluvial Processes and Landforms

1. Meandering streams
   Point bar deposits   Oxbow lakes
   Cutoff meanders

2. Braided streams
   Anastomosing channels Numerous islands

3. Cyclic stream terraces
   Gradient of terrace surface parallel to modern stream
   Cut across rocks of varying resistance

4. Stripped structural surfaces
   Surface parallel to resistant rock layer
   Surface slope has no relationship to modern streams

5. Correlation of cyclic terraces
   Higher terrace remnants are older than lower ones
   Projection of terrace remnants downvalley to match other terrace remnants

6. Drainage type

Drainage patterns	Typical origin
Dendritic	Insequent, consequent
Trellis	Subsequent
Rectanagular	Subsequent
Angular	Subsequent
Radial	Consequent
Annular	Subsequent
Centripetal	Consequent

7. Stream capture by abstraction
   Abandoned channels Beheaded streams
   Wind gaps   Elbows of capture

8. Stream capture by intercision
   Abandoned channels below intersection of meanders

9. Pediments
   Cut across rocks of varying resistance
   Concave-upward profile
   Rock floors

10. Alluvial fans
    Triangular in plan view
    Composed of sediment below the depth of scour of streams

NAME _____

FLUVIAL PROCESSES - CALCULATING RECURRENCE INTERVALS

A recurrence interval (RI) indicates the time recurrence of a given precipitation event. A list of precipitation values are given in Table 5.1. List or rank these according to magnitude, from highest to lowest. The largest magnitude storm will have a RI equal to the number of years of record; the next largest storm will have a RI of years of record divided by two, and so on. Determine the RI according to formula:

$$RI = \frac{n + 1}{m}$$

where n = number of years of record and m = the number of times a given magnitude has been equaled or exceeded, for the 1-hr., 2-hr., and 24-hr. precipitation data. Storms may be ranked according to *classes*, i.e., 1.0" to 1.99"; 2.0" to 2.99", and so on. The RI of each class can then be calculated by tabulating the number of events within each class. Use the table on the next page for your calculations.

Table 5.1 Precipitation Data (Inches per hour)			
YEAR	1-HR	2-HR	24-HR
1935	0.4334	0.4728	0.9456
1936	0.6304	0.6304	2.6398
1937	0.7486	1.3002	2.0488
1938	0.9456	1.0244	2.0488
1939	1.1426	1.2214	2.5216
1940	0.7880	1.0244	2.2064
1941	0.7092	0.7092	1.5760
1942	0.9062	1.1820	2.2064
1943	0.7486	0.7880	1.3396
1944	1.7336	2.2852	3.6642
1945	0.7092	0.7486	3.5066
1946	1.6154	1.6458	2.7186
1947	0.5910	0.7092	2.0882
1948	1.4578	1.8518	2.1276
1949	1.9700	2.3246	2.6398
1950	0.8668	0.8668	2.0488
1951	0.8274	0.9062	1.6548
1952	0.5910	0.5910	1.4184
1953	1.7730	2.2852	2.2852
1954	0.7092	0.7092	2.3246
1955	1.3396	1.4972	3.1126
1956	1.0638	1.4184	2.4880
1957	0.5910	0.7880	2.4034
1958	0.7880	0.9850	1.4578
1959	1.1425	1.4578	2.6792
1960	0.6304	0.6698	1.4184
1961	1.5366	1.7730	2.1276

NAME _____

FLUVIAL PROCESSES - CALCULATING RECURRENCE INTERVALS

Highest amount	Year	Rank	RI	Highest amount	Year	Rank	RI	Highest amount	Year	Rank	RI

FLUVIAL PROCESSES - DISCHARGE RECURRENCE INTERVAL

The recurrence interval (RI) of stream discharge may be calculated in much the same way as the precipitation RI. The table below show the highest flood discharges of a river between 1937 and 1991. Calculate the RI for each discharge and plot it on the vertical axis. From the graph you have drawn determine the expected discharge for a 100-year flood? _____

Table 5.1 Peak Flood Discharges 1935 - 1990 (cfs)				
Discharge	Year	Rank	Discharge	R.I.
21,600	1937			
35,000	1945			
20,000	1947			
24,400	1949			
39,600	1951			
20,000	1953			
28,000	1955			
21,200	1957a			
21,000	1957b			
23,500	1970			
36,000	1975			
24,300	1979			
23,000	1983			
26,000	1984			
29,600	1989			
34,000	1990			

Discharge

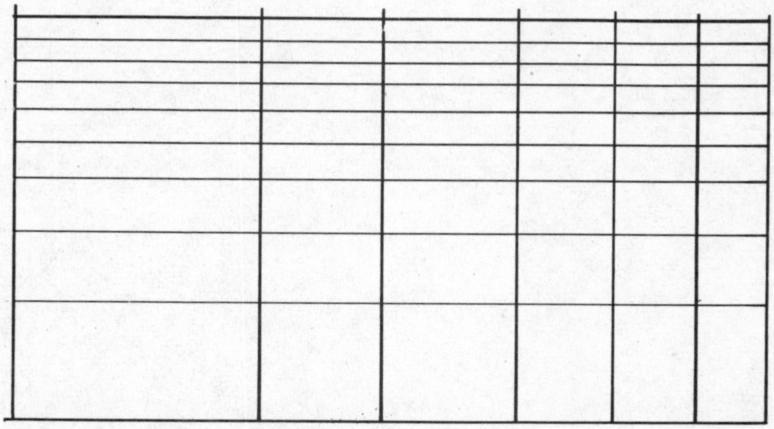

Recurrence Interval, Years

17

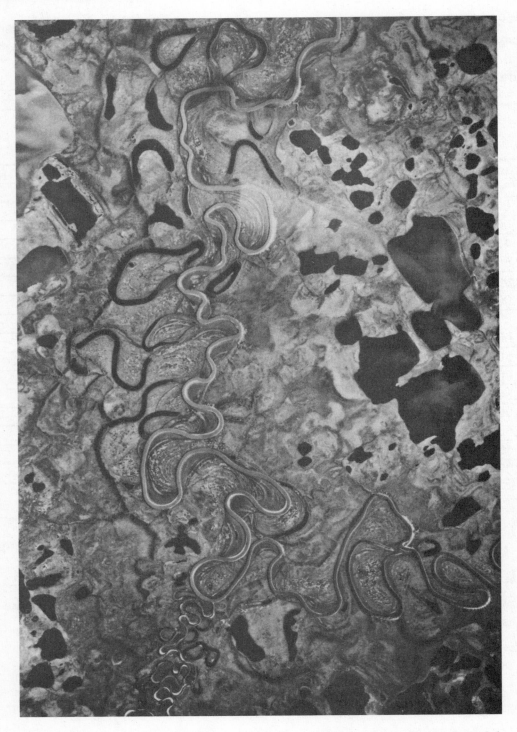

Fluvial landforms make by meandering streams. Note the numerous oxbow lakes and crescentic point bas deposits on the inside of the meanders.

FLUVIAL LANDFORMS – CASPIANA, LOUISIANA

A meandering river pattern is characteristic of rivers crossing low gradient basins underlain by cohesive soil deposits. The channel is narrow and deep.

1. The sinuous, black, dashed line trending NW-SE near the Red River marks the boundary between Caddo and Dossier Parishes (counties). What was the parish boundary based on?

_____

2. What has happened since the parish boundary was established? _____
What is the maximum distance that the Red River has moved since the parish boundary was established? _____

3. Compare the parish boundary with the straight reach of the Red River on the right-hand map. What accounts for the difference in the river course at the time of establishment of the parish boundary and now? _____

4. At Sunny Point (A), the parish boundary goes through part of Sunny Point Lake but not through all of it. Describe the events that must have occurred here between the time of establishment of the parish boundary and now.

    a. _____
    b. _____
    c. _____

5. Name the landform at Sunny Point Lake _____

6. Name the landform at Old River Lake (B). _____

7. Name the landform at Half Moon Lake (B). _____

8. Which is older, Old River Lake or Half Moon Lake? _____
How can you tell? _____

9. Do rivers make good boundaries? _____ Why or why not? _____

10. What are the short, parallel lines along the Red River at (C)? _____ Are they natural or made by humans? _____ How can you tell? _____

11. What name may be given to the type of stream like the Flat River (D) that flows parallel to the Red River for long distances before joining it? _____ Why does this occur?

_____

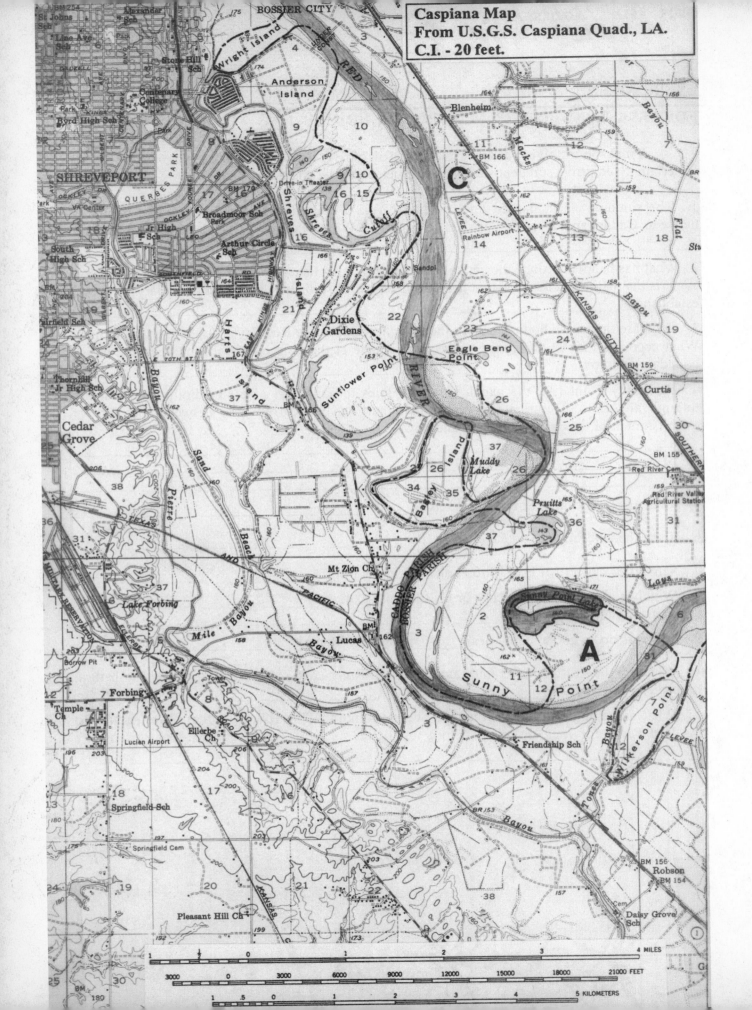

Caspiana Map
From U.S.G.S. Caspiana Quad., LA.
C.I. - 20 feet.

FLUVIAL LANDFORMS – CASPIANA, LOUISIANA

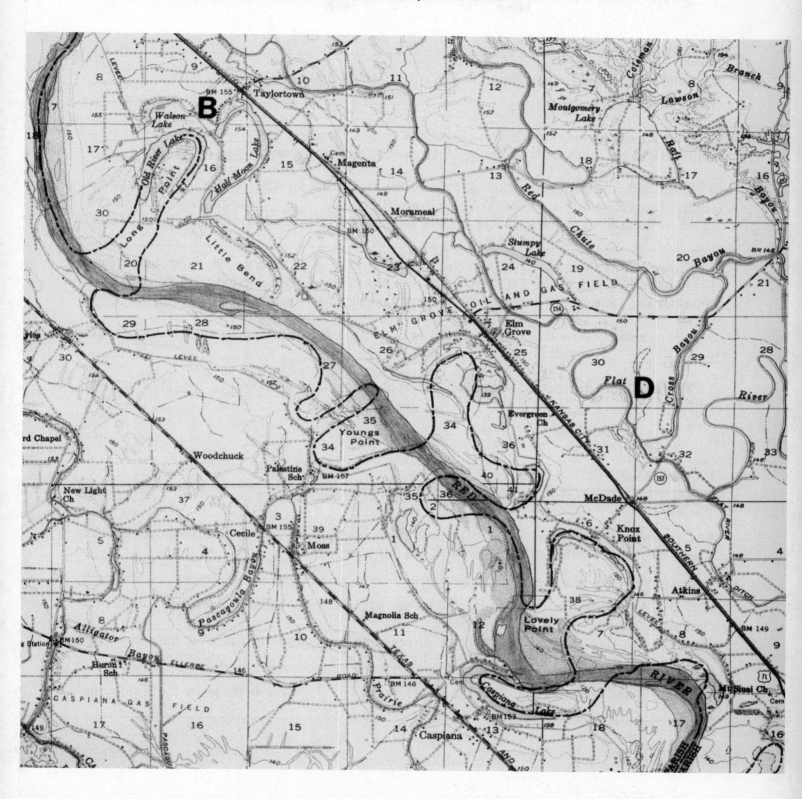

NAME _____

FLUVIAL PROCESSES – SCHLATER, MISSISSIPPI

1. Ashland Brake (A) is a _____

2. Is the surface agent which was responsible for Ashland Brake still operating there? _____
Where has it gone? _____

3. What are the curved ridges and swales east of Ashland Brake? _____
What do they tell you about the history of the stream that made them? _____

4. What is the radius of curvature of the channel at Ashland Brake?_____ What is the
radius of curvature of the channel at Eutah Bend (B)? _____ Considering the similarity
of these two radii of curvature, were Ashland Brake and Eutah Bend made by the same stream?
_____

5. How wide is the channel at Ashland Brake? _____ Compare the width of Ashland
Brake with the width of the Tallahatchie River at Eutah Bend. Considering their difference, do
you think the Tallahatchie River made the topography at Eutah Bend? _____

6. Approximately how far did the channel near Ashland Brake meander before being cut off?
_____ How can you tell? _____

7. Which is older, the curved ridges and swales at (C) or those at (B)? _____ How can you
tell? _____

FLUVIAL PROCESSES – GREENWOOD, MISSISSIPPI

1. Compare the size of the meanders on the Tallahatchie River (A) with those of the Yalobusha
River (B). What conclusion can you draw from this comparison? _____
_____

2. The City of Greenwood (C) is located on the Yazoo River. Is future river transportation to
Greenwood at risk? _____ Why? _____
How can this risk be alleviated to allow Greenwood to continue to use river transportation?
_____

3. What is the origin of Old Orchard Lake? _____ Was it made by the Yalobusha
River? _____ How was it made? _____

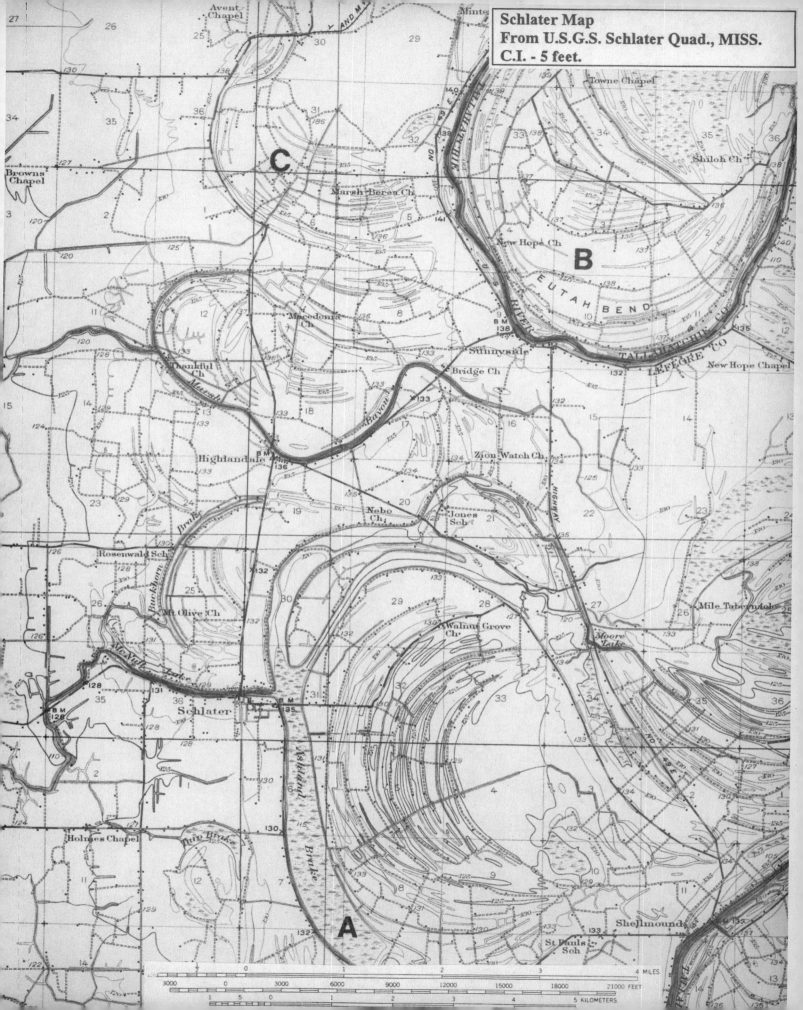

Schlater Map
From U.S.G.S. Schlater Quad., MISS.
C.I. - 5 feet.

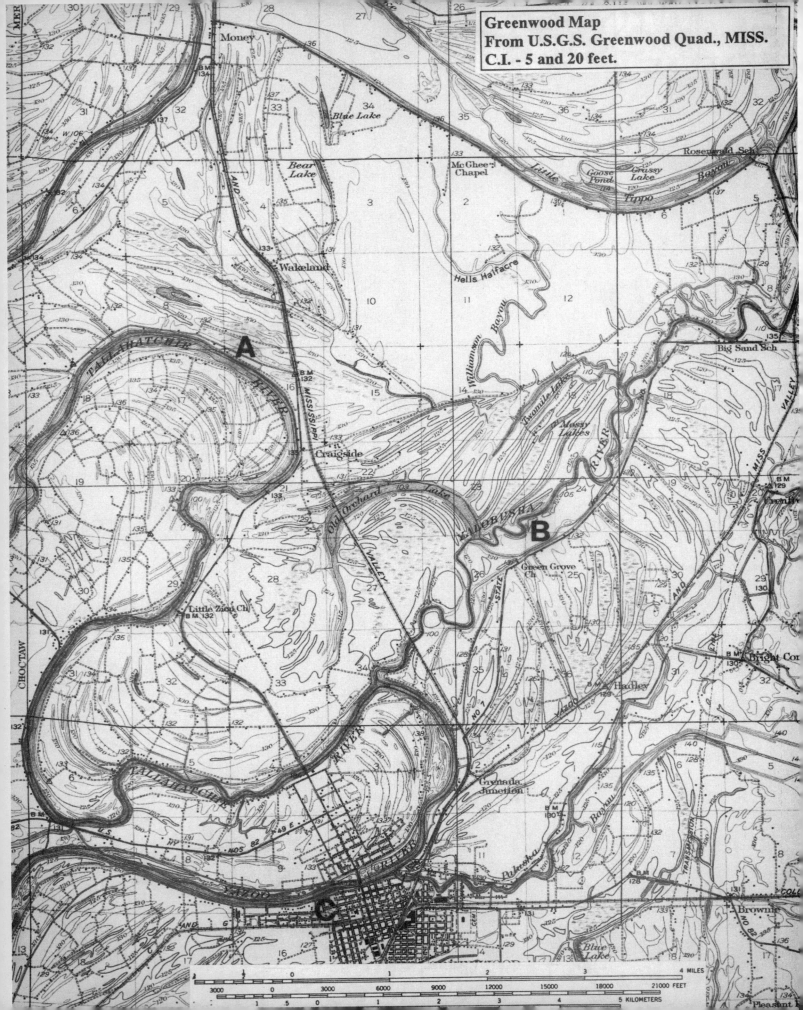

Greenwood Map
From U.S.G.S. Greenwood Quad., MISS.
C.I. - 5 and 20 feet.

FLUVIAL PROCESSES – MISSISSIPPI

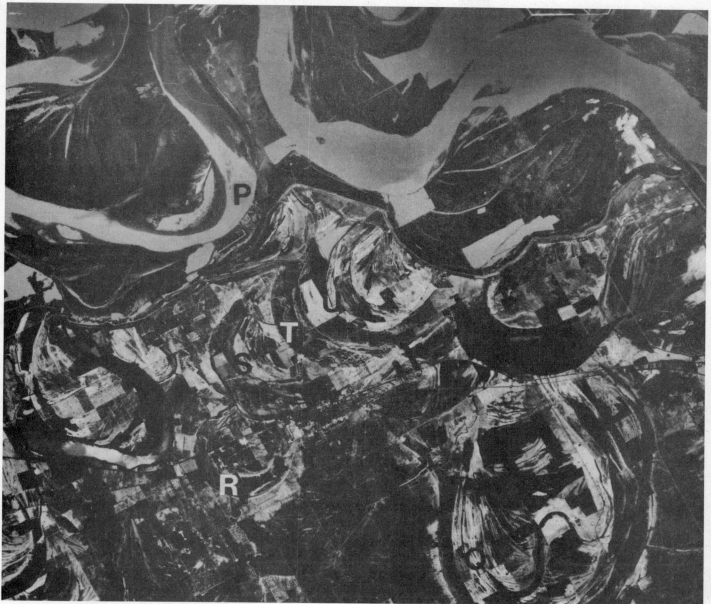

1. Why do streams meander? _____

2. Describe the sequence of events leading to the development of the lake at (P)

   _____

3. What name may be applied to such a lake? _____

4. What are the curved ridges and swales at (Q)? _____

5. Note the curved topography at R, S, T, and U. Describe the sequence of events leading to the development of these features. _____

List them in the order of their ages (oldest) _____ _____   _____ _____ (youngest)

NAME _____

FLUVIAL PROCESSES – BELLEVUE, IDAHO

1. List the characteristics of the meandering stream shown in the photo.

a. _____

b. _____

c. _____

2. List the characteristics of the braided stream shown in the photo.

a. _____

b. _____

c. _____

Which stream has a wide shallow channel? _____

Bellevue, Idaho

3. The segment of the channel of the Big Wood River from 'A' to 'B' is meandering but the river becomes braided from 'B' to 'C.' Downstream from 'C' the river resumes a meandering course. The gradient of most streams becomes more gentle downstream. Is that true of the Big Wood River? Compare the gradient of segment 'A-B' with that of segment 'B-C'.
A-B _____   B-C _____

4. Knowing that braided streams are typically wider and shallower than comparable meandering streams, can you suggest a reason for the downstreams increase in gradient of the Big Wood River? _____

Is this what would be predicted from the Chezy equation ($v = \sqrt{c}\ RS$)? _____ Why?

_____

5. A common misconception about braided streams is that they are "overloaded" so deposit sediment on the floor of the channel, resulting in the formation of many islands and leading to a braided pattern. Could that be the case with the Big Wood River? _____ Why or why not?

_____

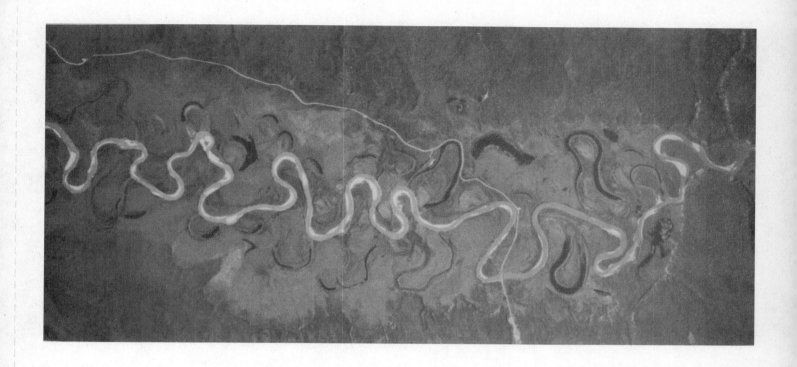

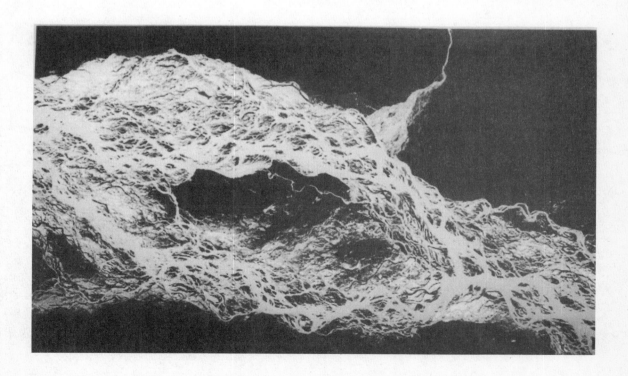

FLUVIAL PROCESSES – WEST BROWNSVILLE AND
LA PALOMA, TEXAS

Two editions of these maps are available, 1936 and 1970, allowing comparisons of changes in features over a 34 year period. The area is part of the floodplain of the Rio Grande River which makes the boundary between the U. S. and Mexico.

Questions 1-6 refer to the l936 edition.

1. Note the small contour interval of the 1936 maps (1 foot). What is the relief? _____

2. How did Trevino Canales Banco no. 5 (A) form? _____

3. How did Matamoros Banco no. 121 (B) form? _____ How can you tell the relative age of Trevino Canales Banco no. 5 and Matamoros Banco no 121? _____ Which is older? _____

4. Where is the stream that once occupied the channel of Resaca del Rancho (C)? _____ Could the Rio Grande have once occupied this channel? _____ The channel is too long to have been abandoned due to a meander cutoff. How else could such an extensive reach of the stream channel be abandoned? _____

5. Draw a topographic profile along the line D-D'. How do you account for the elevation of the channel above the surrounding floodplain? _____

6.   What can happen to the channel if the levees are breached during a flood? _____

7.   What is the basis for the international boundary between the U.S. and Mexico? _____ Was this a wise choice? _____ Why or why not? _____

8. Compare the 1936 edition of the West Brownsville map with the 1970 edition and note the changes that have occurred in the course of the Rio Grande during the period between maps. What has happened at 'E'? _____ What has happened at 'F'? _____

9. Which country owned the land at 'E' on the 1936 map? _____ Who does it belong to in 1970? _____. Who owned the land at 'F' on the 1936 map? _____ Who does it belong to in 1970? _____

10. How did the crescentic topographic at 'G' form? _____ When did this occur? _____

11. How far has the meander at 'H' moved since 1936? _____ The meander at 'I'? _____ What has happened at 'J' since 1936? _____

FLUVIAL PROCESSES – WEST BROWNSVILLE MAP (1936)

FLUVIAL PROCESSES – WEST BROWNSVILLE MAP (1970)

FLUVIAL PROCESSES – LA PALOMA, TEXAS MAP (1970)

FLUVIAL LANDFORMS – GRASS CREEK BASIN, WYOMING

1. Note Wagonhound Bench (A).

    (a) What is its *height* above the present streams? _____

    (b) What is its *direction* of slope? _____

    (c) What is its *gradient*? _____ ft/mile

2. Draw topographic profiles of Wagonhound Bench along line A-A' and Cottonwood Cr. along line B-B'.

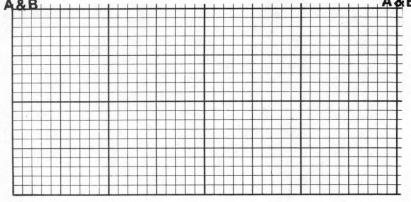

3. Is Wagonhound Bench a cyclic or non-cyclic terrace? _____ What evidence *from the map* supports your conclusion? _____

4. Describe how Wagonhound Bench formed? _____

5. The 5500' contour at 'C' makes a right-angle bend at 5508'. What is the reason for this?

_____

Draw a profile along the line D-D'.

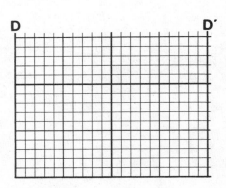

How high is the SE bench above Wagonhound Bench? _____ What is its origin?

_____

Which is older, the higher or lower bench? _____How do you know?

_____

6. The broad surface near 'E' abutts a scarp (shown by two closely-spaced contours). What is the reason for this? _____

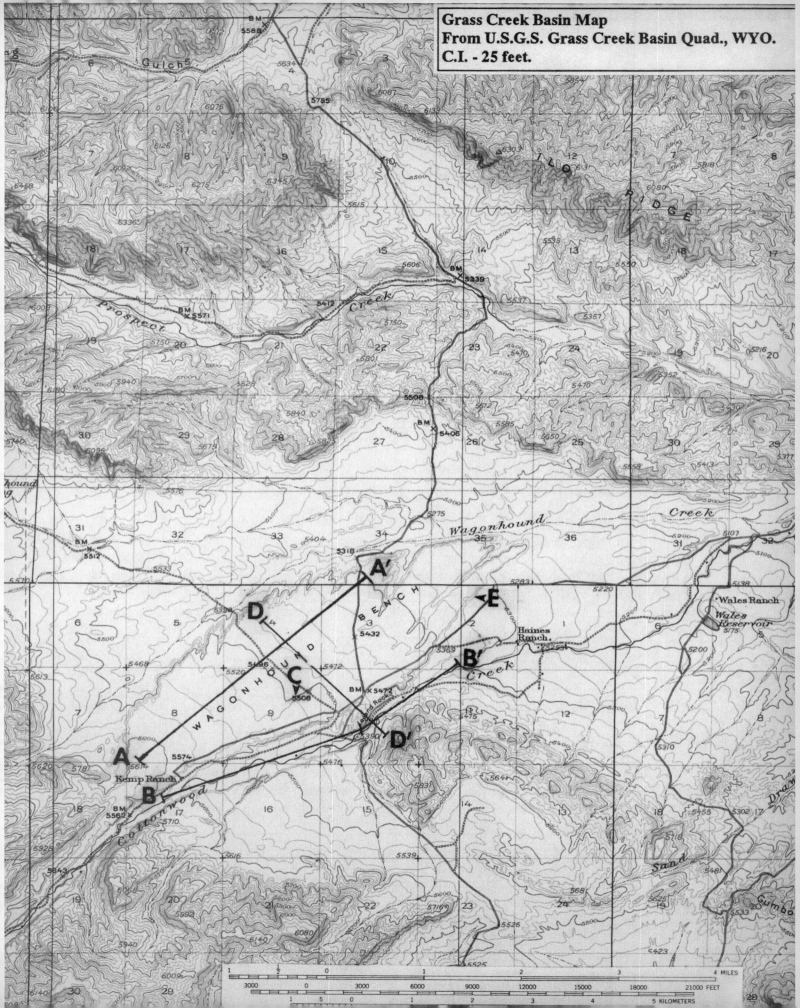

Grass Creek Basin Map
From U.S.G.S. Grass Creek Basin Quad., WYO.
C.I. - 25 feet.

FLUVIAL LANDFORMS – MEETEETSE, WYOMING

1. Draw topographic profiles along the lines A-A; and B-B' on the two benches that border the Wood River on the west and along line C-C' on the present floodplain of the Wood River.

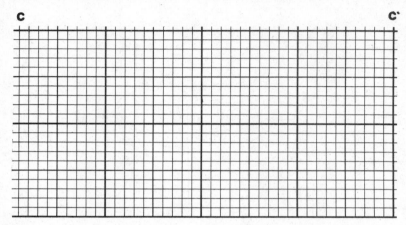

2. Compare the gradient of the benches with that of the present Wood River (ft/mile).
Bench at A-A' _____ Bench at B-B' _____ Wood River _____

3. Based on the gradients of the benches and the present Wood River, are the benches cyclic or non-cyclic? _____

4. Which is older, the terrace at A-A' or the terrace at B-B'? _____ How do you know? _____

5. Compare the gradient of the bench at *'C'* with the gradient of the present Greybull River. Does it suggest that the bench is cyclic or non-cyclic? _____ How did the bench form? _____ Projection of the profiles of terraces A-A' and B-B' along the Wood River suggests that the bench at *'C'* correlates with terrace _____

6. Why doesn't bench *'C'* extend downstream along the rest of the Greybull River? _____
_____

7. Note the broad, gently-sloping surfaces at *Meeteetse Rim* (D) that are deeply incised by Long Hollow and Meeteetse Creek. Is this surface cyclic or non-cyclic? _____ What evidence from the map supports your conclusion? _____

8. Draw a topographic profile along the surface at D. Project the profile eastward to the flat surface at 'F'. Could the surface at Meeteetse Rim have been graded to the surface at 'F'? _____
What   evidence   from   the   map   supports   your   conclusion?
_____

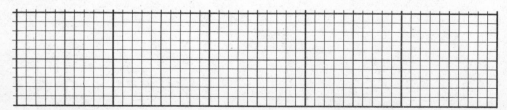

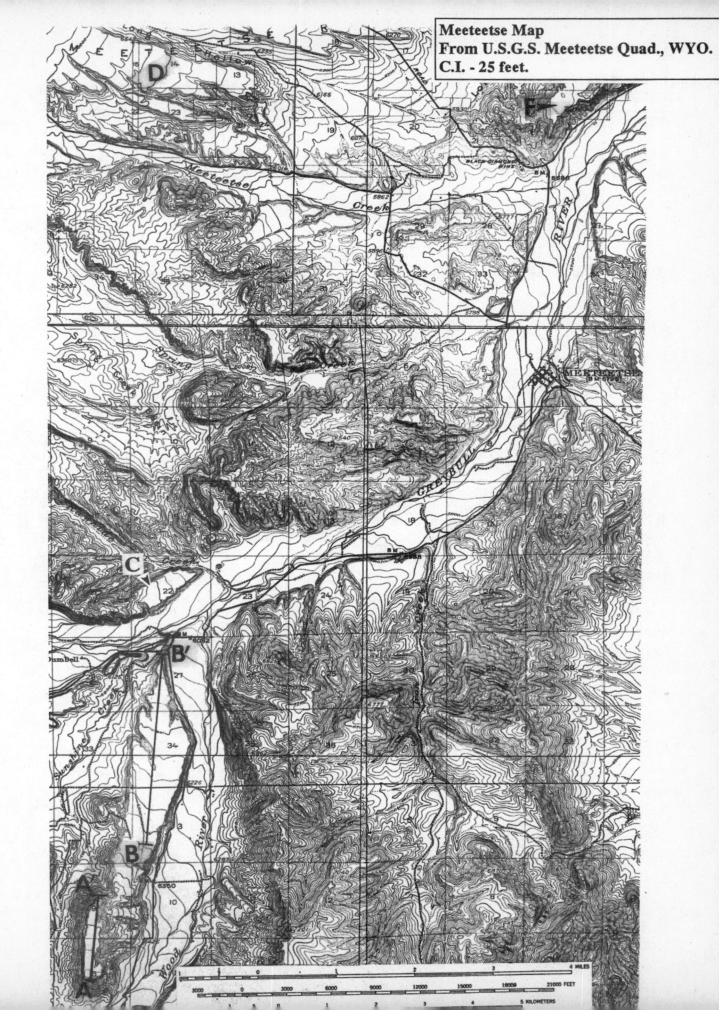

Meeteetse Map
From U.S.G.S. Meeteetse Quad., WYO.
C.I. - 25 feet.

FLUVIAL LANDFORMS – YU BENCH, BURLINGTON, OTTO, WYO.

YU bench, Emblem Bench, and Table Mt. are cut across the upturned edges of tilted Tertiary sandstone and shale and truncate rocks of varying resistance.

1. Draw a topographic profile along the line A-A' on Yu Bench.

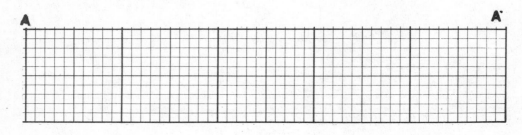

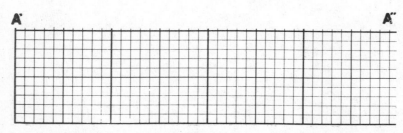

2. On the same profile, draw a profile of the present valley of the Greybull River along the line B-B'. Compare the gradient of the present Greybull River with YU Bench. What is the slope of the present Greybull (in feet per mile)? _____ What is the slope of YU Bench (in feet per mile)? _____ Is YU Bench a cyclic erosion surface? _____ Is this consistent with the fact that the benches truncate bedrock structures? _____

_____

3. Draw a topographic profile along line C-C' on Emblem Bench. Compare the gradient of Emblem Bench with that of YU Bench and the Greybull River. Are they similar? _____ Is Emblem Bench a cyclic erosion surface? _____

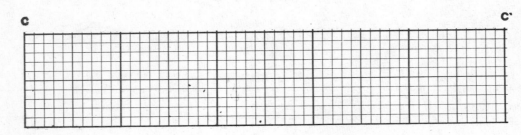

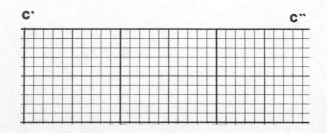

FLUVIAL LANDFORMS – YU BENCH, BURLINGTON, OTTO, WYO.

Which is older, the Yu Bench surface or the Emblem Bench surface? _____
How do you know? _____

5. The contours that cross Emblem Bench make a sharp turn midway across the bench (at 'D').
What does this indicate about the nature of the surface of the bench? _____
_____

Which side of Emblem Bench is older, C-C' or 'D'? _____ Why? _____
_____

6. Draw a topographic profile along the line E-E' on Table Mt. Compare the gradient of the
surface of Table Mt. to that of Yu Bench, Emblem Bench, and the modern Greybull River. Are
they similar? _____

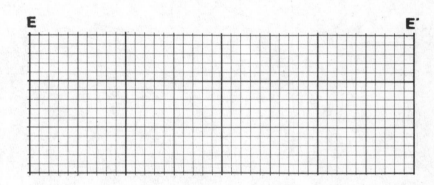

7. Project the gradient of Yu Bench and Emblem Bench to Table Mt. (E-E'). Does the projected
profile of the Emblem Bench surface match up with the Table Mt. surface? _____ Does the
projected profile of the Yu Bench surface match up with the Table Mt. surface?
_____

8. How much downcutting has taken place between the Table Mt. terrace and the present
floodplain of the Greybull River. _____

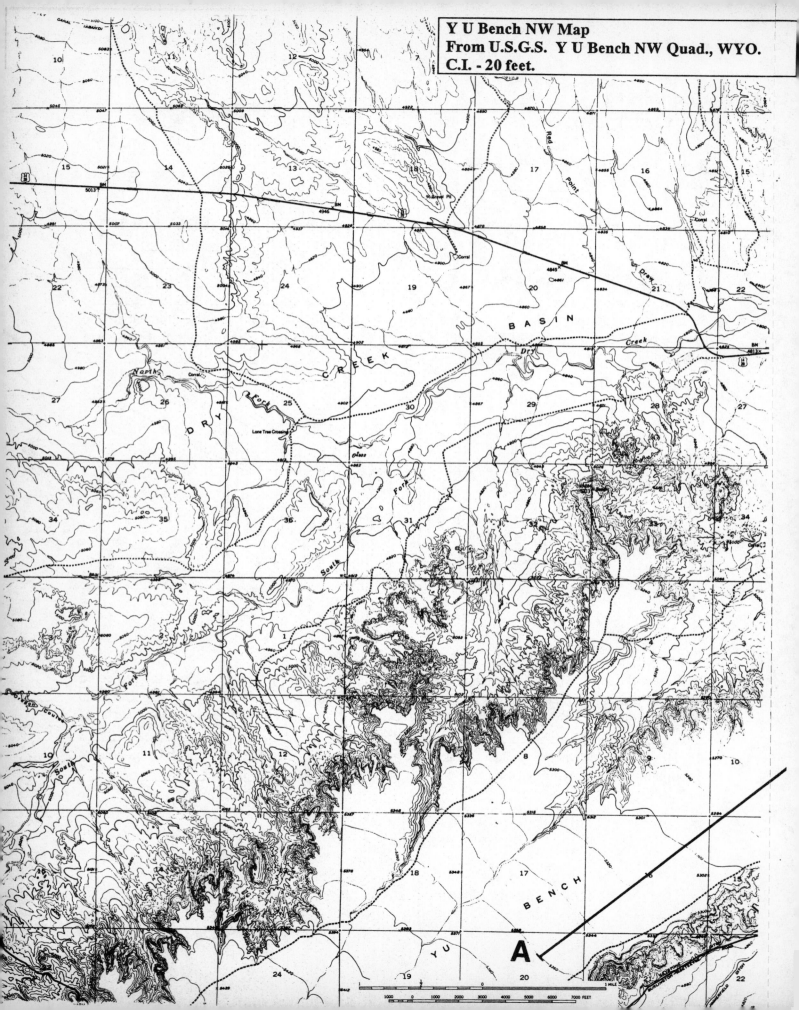

Y U Bench NW Map
From U.S.G.S. Y U Bench NW Quad., WYO.
C.I. - 20 feet.

A

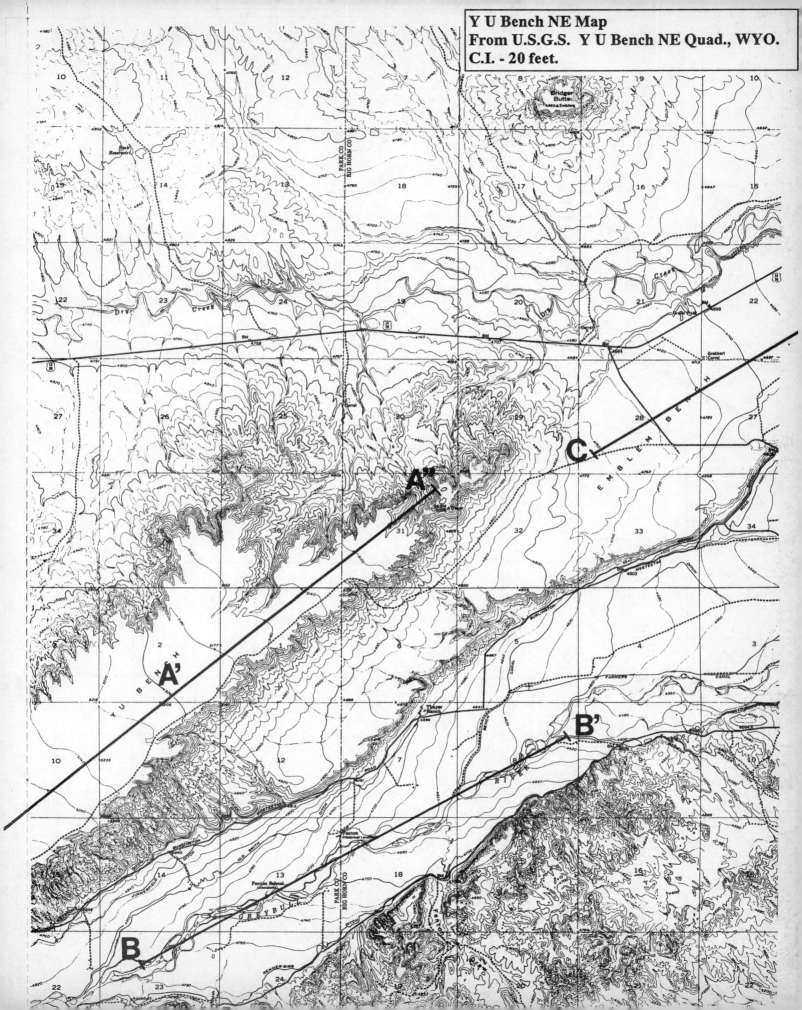

Y U Bench NE Map
From U.S.G.S. Y U Bench NE Quad., WYO.
C.I. - 20 feet.

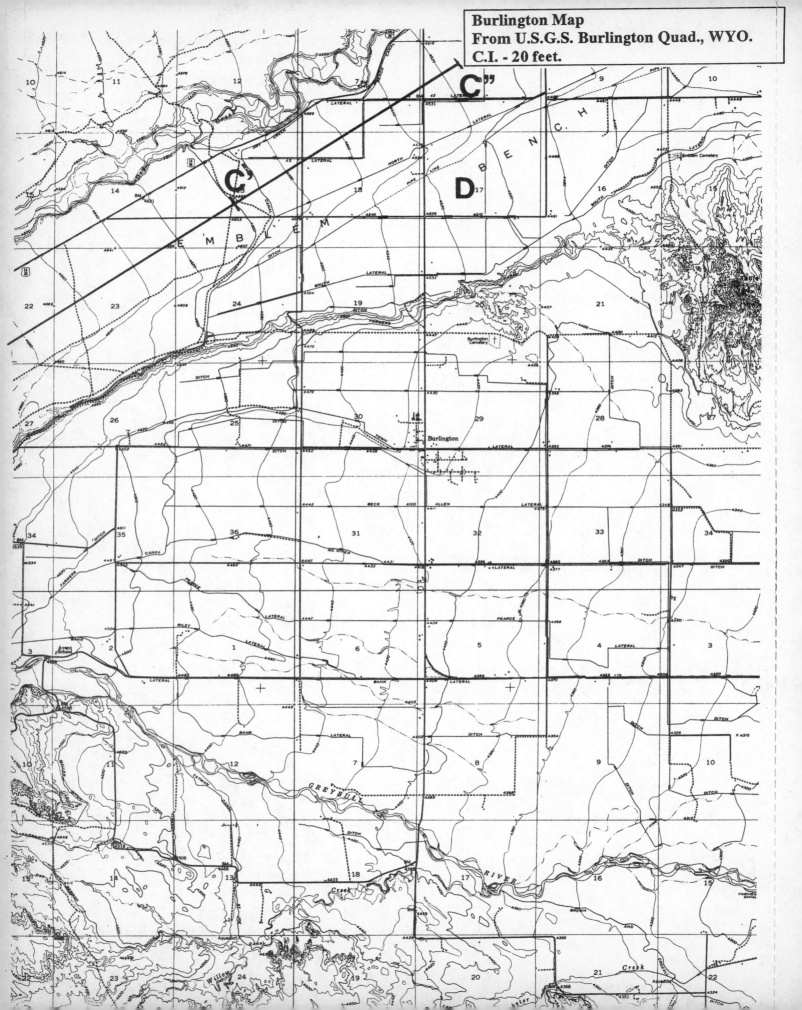

Burlington Map
From U.S.G.S. Burlington Quad., WYO.
C.I. - 20 feet.

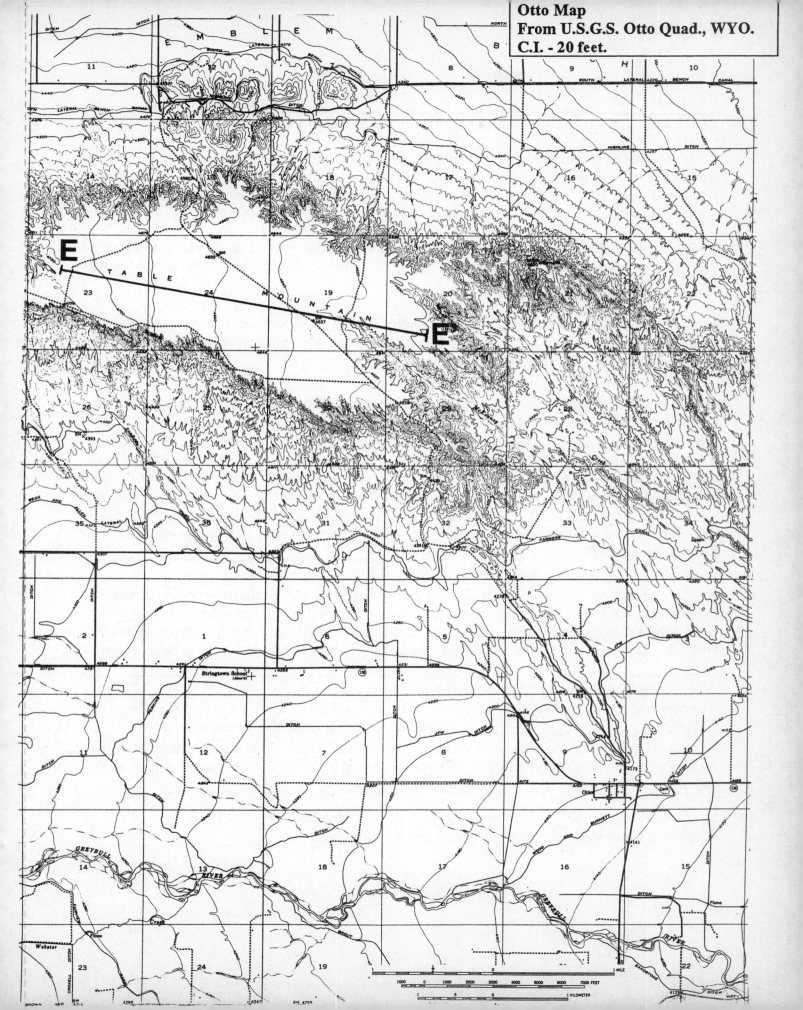

FLUVIAL LANDFORMS – LITTLE MUDDY, KENTUCKY

Here is an exercise that illustrates drainage changes and will challenge your command of the method of multiple working hypotheses. The streams that play central roles in the drainage changes all have very low gradients, as shown by the general absence of contours that cross the streams. The Green River flows westward and the Barren River flows northward.

1. Black Swamp Branch (A) is a minor, intermittent stream that lies in a broad valley and seems too small to have eroded the valley in which it now flows. Such a stream is known as an _____ stream.

2. The southward continuation of that same valley is now occupied by the lower part of Little Muddy Creek (B). Could Little Muddy Creek have made the valley now occupied by Black Swamp Branch? _____ Why or why not? _____

3. The only other streams in the area that could have made the valley of Black Swamp Branch are the Barren River and the Green River.

If the Barren River did make the valley (A-B), you are then faced with the problem of how to get it out of that valley and into its present valley (C). Could the Barren have made the valleys at B and C simultaneously or intermittently? _____ Could a meander of the Barren have flowed up the valley at 'C' until it intersected the Green River? _____ Why or why not? _____

Therefore, what can you conclude about the role of the Barren River in making the valley of Black Swamp Branch (A)? _____

4. Are the valleys of the Green River and the valley at A and B similar enough in size to consider that the Green River may have once flowed through Black Swamp Branch? _____
If so, draw a sketch map showing the position of the Barren River and Little Muddy Creek at that time. If the Green River ever flowed in that valley, how could it get to its present course?

_____
Where would the Barren River have flowed in that circumstance? _____
Sketch a map on a separate piece of paper. Where would Little Muddy Creek have flowed in that case? _____ Show this on the sketch of the Barren River's course.

How could the Barren have gotten into its present course? (Hint: Consider the flatness of the valley floors, the height of the *surface* of the river during high flood stages, and the most likely path that the river would take) _____

_____

How could Little Muddy Creek have gotten into its present course? _____

Little Muddy Map
From U.S.G.S. Little Muddy Quad., KY.
C.I. - 20 feet.

FLUVIAL LANDFORMS – FRANKFORT, KENTUCKY

The large stream on the map is the north-flowing Kentucky River, which is joined from the west by Benson Creek (B) at the city of Frankfort.

1. Note the large, incised, abandoned meander (A) that extends eastward from Frankfort through Thorn Hill before bending westward to the Kentucky River. Could this channel have been cut by the drainage that now occupies it? _____ Why or why not?
_____

2. What are the only two streams large enough to have cut this channel?
_____   _____

3. From your answer to question 2, the problem now becomes: how did whichever of the two streams that must have cut this channel get out of that channel and into the present channel. In attempting to solve this problem, the method of multiple working hypotheses leads us to consider several possibilities:

(a) Is the present valley of Benson Creek (B) large enough to suggest that Benson Creek could be capable of cutting the abandoned valley? _____

(b) Could Benson Creek have cut the abandoned valley while the Kentucky was in its present position? _____ Why or why not? _____

(c) Your conclusion to the previous question leaves us with the Kentucky River as the only stream that could have cut the abandoned valley. If that is the case, we must then answer the question: if the Kentucky River once occupied the abandoned valley, how did it leave the valley to its present northward course from Frankfort (C)? Could it have been caused by a meander cutoff of the Kentucky River? _____ Why or why not? _____
_____

This leads us to the question of how the *valley* of the Kentucky River north of Frankfort originated.

In attempting to answer this last question, a clearer picture of the situation may be gained by reconstructing the drainage at the time the Kentucky River must have occupied the abandoned valley. At this time, the only stream that could have made the valley north of Frankfort was _____ How then could the Kentucky River have been diverted through this valley? _____ What is this process called? _____

Frankfort, KY. Map
From U.S.G.S. Frankfort East and West Quad., KY.
C.I. - 10 feet.

CONTOUR INTERVAL 10 FEET

FLUVIAL LANDFORMS – GILA BUTTE, ARIZONA

1. Draw a profile along line A-A'.

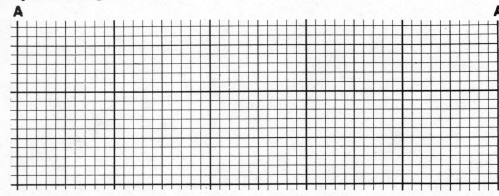

2.   What is the gradient of the gently-sloping surface in sect. 36 along line A-A'? _____ ft/mile. What is the gradient of the surface near sect. 13 along line A-A'? _____ ft/mile. Graded streams have concave-upward profiles. Is the surface along A-A' concave upward? _____. What conclusion can you draw from your profile?
_____

3. What name may be given to the concave-upward, gently-sloping surface along line A-A'?
_____

4. This surface is developed on a thin veneer of alluvial sand and gravel and cuts across rocks of varying resistance, so it must be _____ (erosional or depositional) in origin.

5.   This surface differs from a normal floodplain in that no large, single stream of any consequence flows across it. What then did produce this broad, gently-sloping surface?
_____

6. The isolated rock knobs (B) that rise above the gently-sloping surface are erosional remnants of formerly more extensive rock masses. What name may be applied to them? _____ Why have they not been consumed by erosion? _____
_____

7. What name may be applied to the gap in the Sacaton Mts. at 'C'? _____

8. What will eventually happen to the Sacaton Mts.? _____

9. How might you distinguish a pediment from an alluvial fan or bajada?
_____

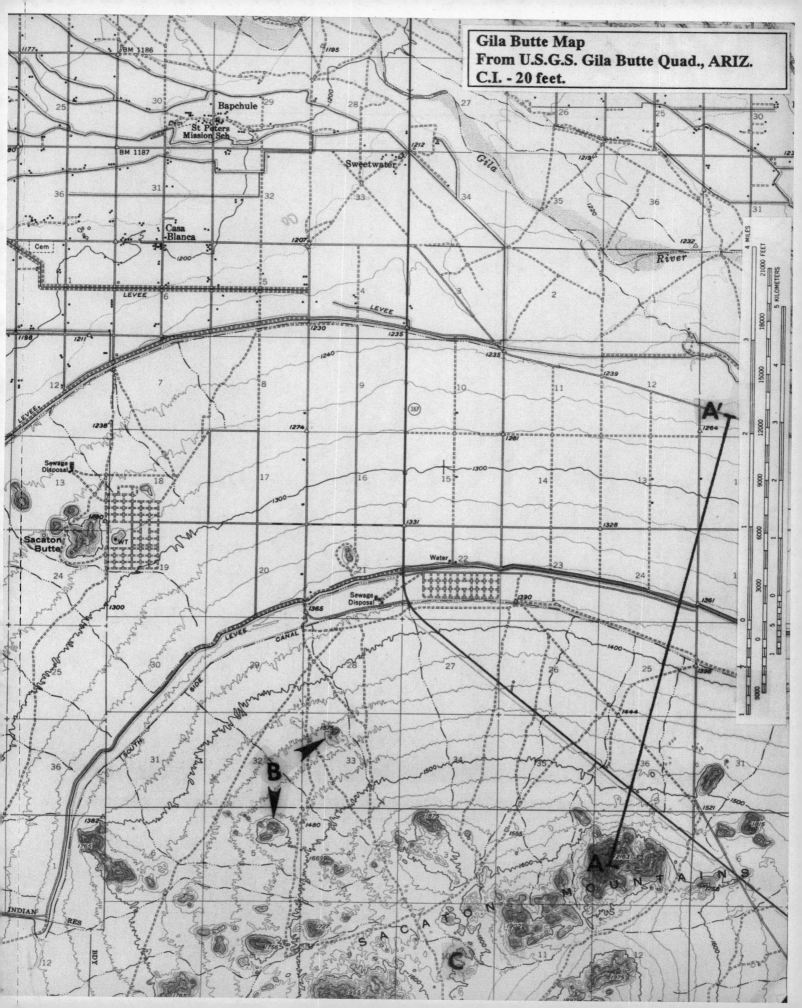

Gila Butte Map
From U.S.G.S. Gila Butte Quad., ARIZ.
C.I. - 20 feet.

FLUVIAL LANDFORMS

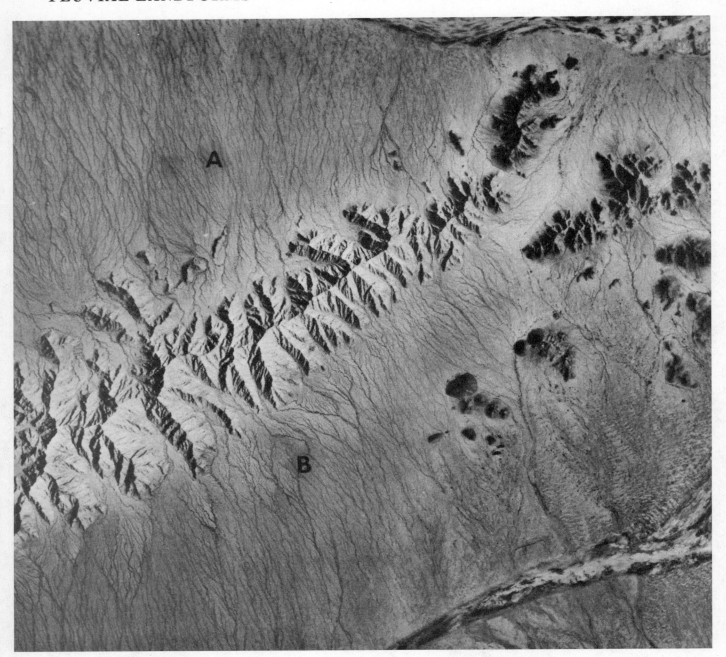

Compare the air photos with the Gila Butte topographic map. Although the air photos come from a different area than the map, note the similarity of geomorphic forms.

1. The gently-sloping surfaces at 'A', 'B', and 'C' are of the same origin as those on the Gila Butte map. Note that no single, large stream crosses any of the surfaces. Instead, hundreds of small gullies can be seen on the air photos.

2. What name may be applied to the rock knobs at 'D' _____

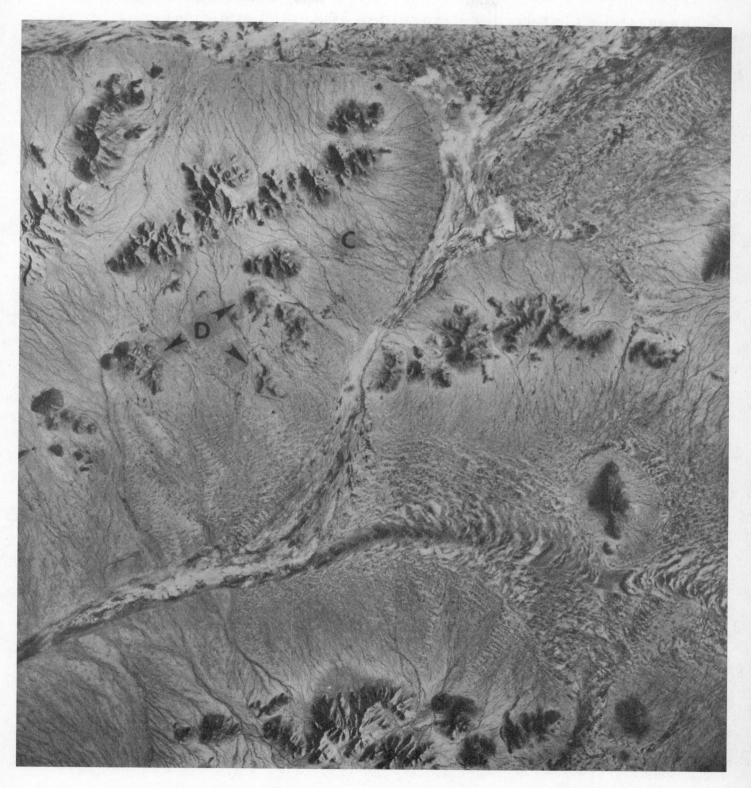

Groundwater

Topographic Criteria for Recognition of Karst Landforms

1. Sinkholes - irregular depression, often filled with water.

2. Sinking streams - lack of an integrated surface drainage. Rivers and creeks may be intermittent, with many dry valleys with streams that disappear and reappear.

3. Solution valleys

4. Blind valleys

5. Caves

6. Karst windows and bridges

7. Cockpit karst

Karst terrain in northern Florida. The dark, oval lakes fill sinkholes in limestone.

GROUNDWATER - PARK CITY AND SMITH'S GROVE, KENTUCKY

1. The numerous depressions on the Smith's Grove map have formed by _____
of _____. What name may be applied to the depressions? _____
What kind of rock occurs there? _____

2. How big are the depressions (on the average)? width _____ depth _____

3. What are the two ways in which these depressions may form?
(a)_____ (b) _____

4. Little Sinking Creek flows westward from the east edge of the Park City map for about 6 miles (10 km) before terminating abruptly at 'A'. What happens to the creek at its distal end?

_____
Why does this occur? _____
What name is given to this kind of _creek_? _____
What name is given to the _valley_? _____

5. Two other creeks on the maps show the same characteristics. Name the creeks.
(a) _____ (b) _____

6 Except for the northernmost part, the Park City map is relatively free of depressions. What is the reason for their absence? _____

7. Pilot Knob ('B' on the Smith's Grove map) rises about 250 feet above the level of the depression topography that surrounds it. Because it stands as a high area, it must be made of more resistant rock than the adjacent area. What kind of rock makes up Pilot Knob? _____ How thick must this rock layer be? _____
Is it older or younger than the rock making up the depression topography? _____

8. Draw a topographic profile and geologic cross section along the line C-C,' showing the relationship between the topography and the geology.

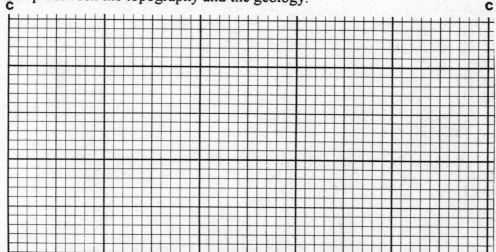

9. How can you distinguish between karst topography and kame-kettle topography on a topographic map without knowing the geology?

_____

Smiths Grove Map
From U.S.G.S. Smiths Grove Quad., KY.
C.I. - 10 feet.

Park City Map
From U.S.G.S. Park City Quad., KY.
C.I. - 10 feet.

GROUNDWATER – LAKE WALES, FLORIDA

1. What is the origin of the many lakes on this map? _____

2. What kind of rock occurs here? _____

3. Some of the depressions on this map are filled with lakes, others are not. What are the two possible reasons for the lakes that occupy depressions?

(a) _____ (b) _____

4. The elevations of many of the lakes (shown by the numbers on the lakes) indicate a systematic change from one part of the map to the other. Assuming that the elevation of these lakes approximates the level of the water table in the area, draw contours on the map using a 2-foot contour interval to show the water table.

EXERCISE 7.2

GROUNDWATER – CRYSTAL LAKE, FLORIDA

1. What is the origin of the many lakes on this map? _____

2. How wide are the larger lakes on this map? _____

3. How deep below the general level of topography are these lakes? _____

4. The elevation of many of the lakes shows a systematic change from one part of the map to the other. Assuming that the elevation of these lakes approximates the level of the water table in the area, draw contours on the map using a 5-foot contour interval to show the water table.

Lake Wales Map
From U.S.G.S. Lake Wales Quad., FA.
C.I. - 5 feet.

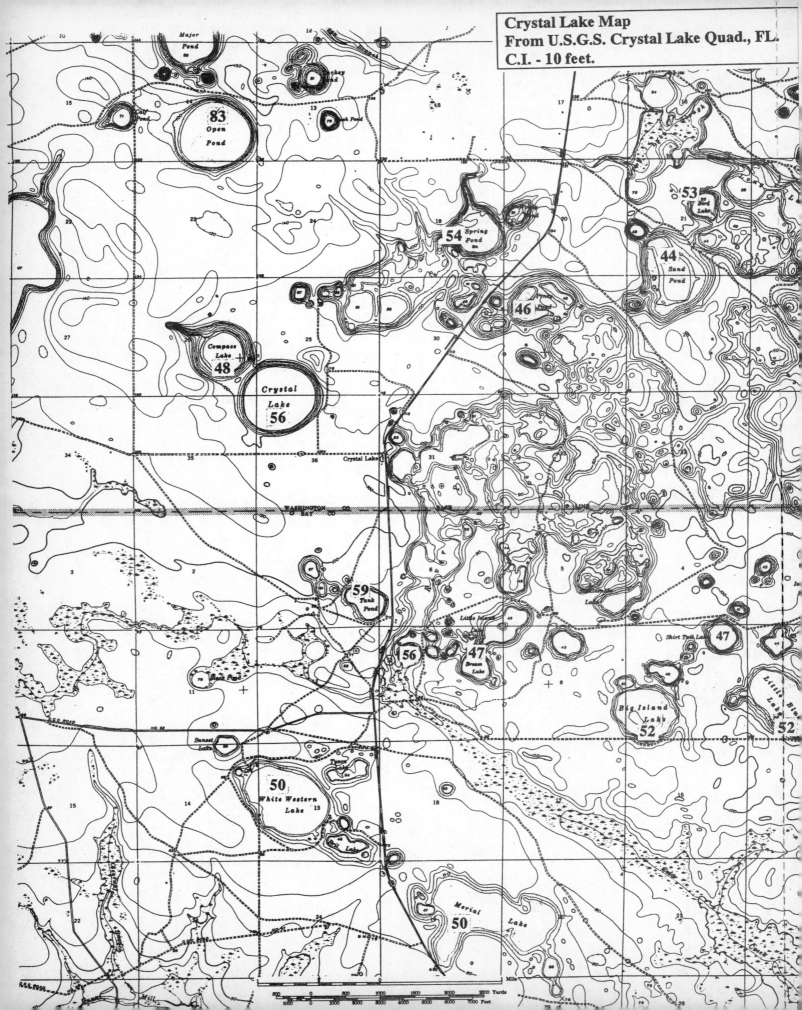

Crystal Lake Map
From U.S.G.S. Crystal Lake Quad., FL.
C.I. - 10 feet.

Topographic Expression of Folded Strata

Criteria for Recognition of Sedimentary Rocks

1. Lithology
   a. Fine-grained clastic rocks (shale, siltstone)
      (1) Tend to have dark photographic tones
      (2) Fine-textured drainage
      (3) Joints not usually present (or closely-spaced)
      (4) Differential erosion to form valleys
   b. Coarse-grained clastic rocks (sandstone, conglomerate)
      (1) Tend to have light photographic tones
      (2) Coarse textured drainage
      (3) Usually conspicuously jointed, wide-spaced joints make distinctive topography.
      (4) Differential erosion to form ridges
      (5) Commonly large talus blocks at base of cliffs.

2. Structure
   a. Flat-lying beds - cliff and bench topography, plateaus, mesa, buttes
   b. Tilted beds - ridge and valley topography
      (1) Homoclinal ridges parallel if axis is nonplunging; zigzag if plunging
      (2) Asymmetry of homoclinal ridges
      (3) V-shaped notches through homoclinal ridges.
      (4) Flatirons
      (5) Tributary streams are longer flowing on down dip slopes than scarp faces
   c. Folds
      (1) Anticlines have gently tapering noses
      (2) Synclines have short, blunt noses
      (3) Notches through anticlinal ridges may be inverted (U-shape). Contours may appear to be offset at cliff face.

TOPOGRAPHIC EXPRESSION OF FOLDED STRATA –
WOLF POINT, WYOMING

1. Are the rocks here sedimentary or crystalline? _____ What is the evidence
for your answer? _____
What kind of rock makes the ridges? _____ The valleys? _____

2. The apex of the V-shaped notches where Willow Creek crosses the ridge at 'A' points in the
direction of _____ Thus, the bed that makes the ridge at 'A' dips to the
_____ and the strike is _____

3. Draw a topographic profile along line C-C'.

Note that the asymmetry of the ridge at 'B' may also be used to determine the direction (the
gentle dip slope is inclined in the direction of dip).   Calculate the angle of dip in degrees
_____

4. What is the drainage pattern on the map? _____ What controls it? _____

5. Red Canyon Rim (D) is a _____ (geologic structure) ridge made of
_____ that strikes _____ and dips _____

6. Red Canyon is a _____ (geologic structure) valley made of
_____ that strikes _____ and dips _____

7. The uniform slope of the surface at 'E' is a stripped structural surface controlled by a resistant
bed that makes a pronounced dip-slope that extends most of the way across the map. However,
the continuity of the dip-slope is interrupted at 'F' as it crosses the asymmetric valley of Barrett
Creek. Each contour is offset sharply to the NE as it crosses Barrett Creek, making a Z-shaped
pattern. What is the most likely reason for the offsetting of contours at Barrett Creek?
_____ Cite the evidence from the map for your answer.

_____

8. Why does Little Popo Agie Canyon become progressively deeper westward? (Hint: compare
the profile of the resistant bed that the river has cut into with the gradient of the Little Popo
Aggie River) _____

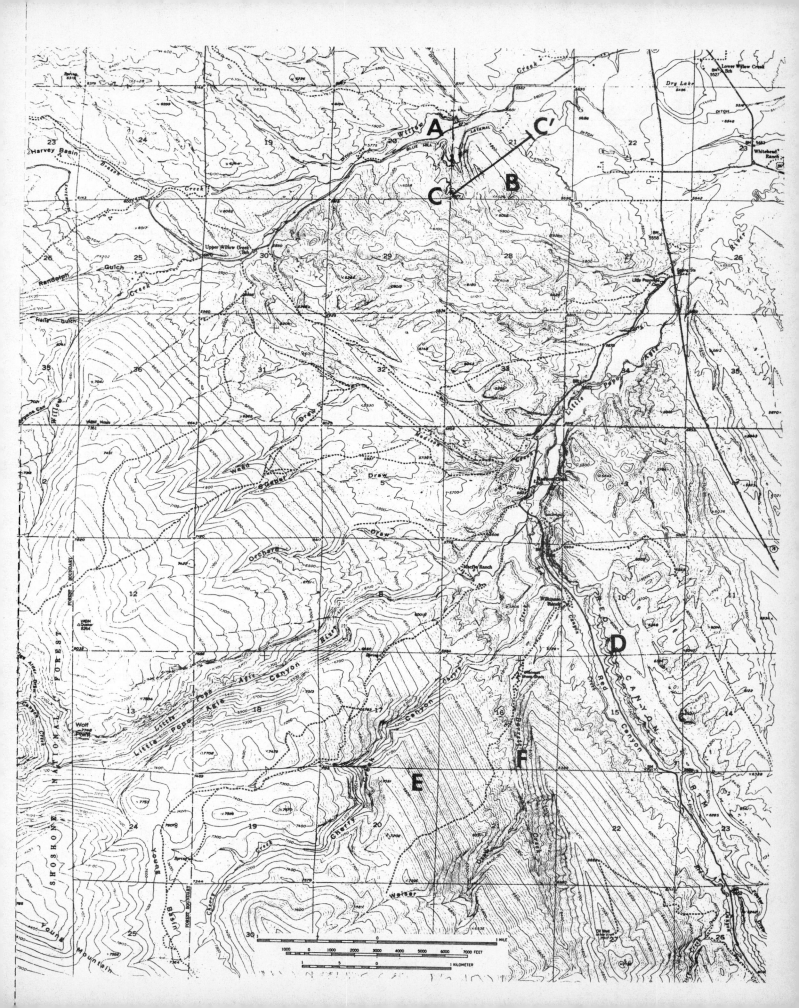

NAME _____

TOPOGRAPHIC EXPRESSION OF FOLDED STRATA –
MILTON AND WILLIAMSPORT, PENNSYLVANIA

The zigzag ridges and valleys on this map indicate folded sedimentary rocks. The geologic structures of this area can be determined using V-shaped notches, asymmetry of ridges, and gently-tapering or blunt nose at the apex of plunging folds. Beds that dip in only one direction make either homoclinal ridges or homoclinal valleys, depending on the resistance of the beds to erosion. Beds that dip in more than one direction make anticlinal ridges, anticlinal valleys, synclinal ridges, or synclinal valleys.

1.   Considering that the climate here is humid, what kind of rock makes up the ridges? _____. What kind of rock makes up the valleys? _____

2. What two topographic features of the ridge at 'A' allow determination of the geologic structure of the ridge? _____ _____

3. What is the direction of dip of the resistant bed that makes the ridge at 'A'? _____

4. Nippenose Valley (B) is surrounded by concentric ridges that indicate it is a breached fold (e.g., Follow the steep scarp of the ridge at 'A' around to the north side of the valley). Draw the axis of the fold from 'B' to 'C', following the crest of the ridge near 'C'. Two lines of evidence allow determination of the structure of this fold: If the bed at 'A' dips to the _____ and may be traced around to the opposite side of Nippenose Valley, what must the structure be? _____. The nose of the structure at 'C' indicates that the structure must be _____ What is direction of plunge of this structure? _____

5. What do Mosquito Valley and Nippenose Valley have in common? _____
What is the high area between them? _____

6.   The nose of the geologic structure at 'D' shows that the geologic structure there is _____ plunging _____

7. Therefore, North White Deer Ridge (E) must be a _____ ridge dipping _____ South White Deer Ridge (F) is a _____ ridge dipping _____, and White Deer Valley (G) is a _____ plunging _____

8. The geologic structure of the ridge at 'H' is _____
The topographic evidence is _____

9. The geologic structure of the ridge at 'I' is _____
The topographic evidence is _____

TOPOGRAPHIC EXPRESSION OF FOLDED STRATA – MILTON AND WILLIAMSPORT, PENNSYLVANIA

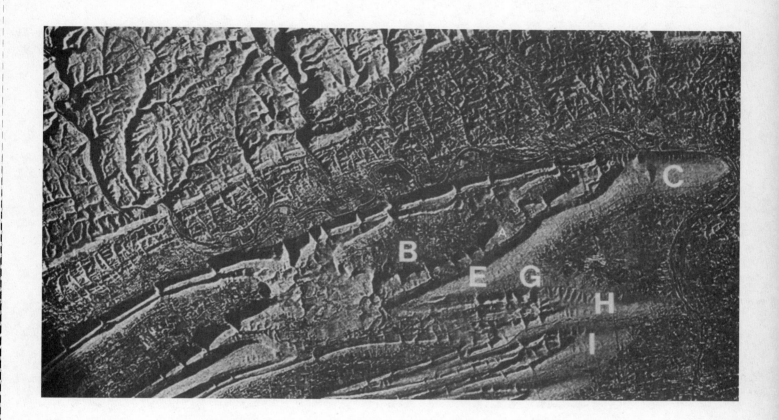

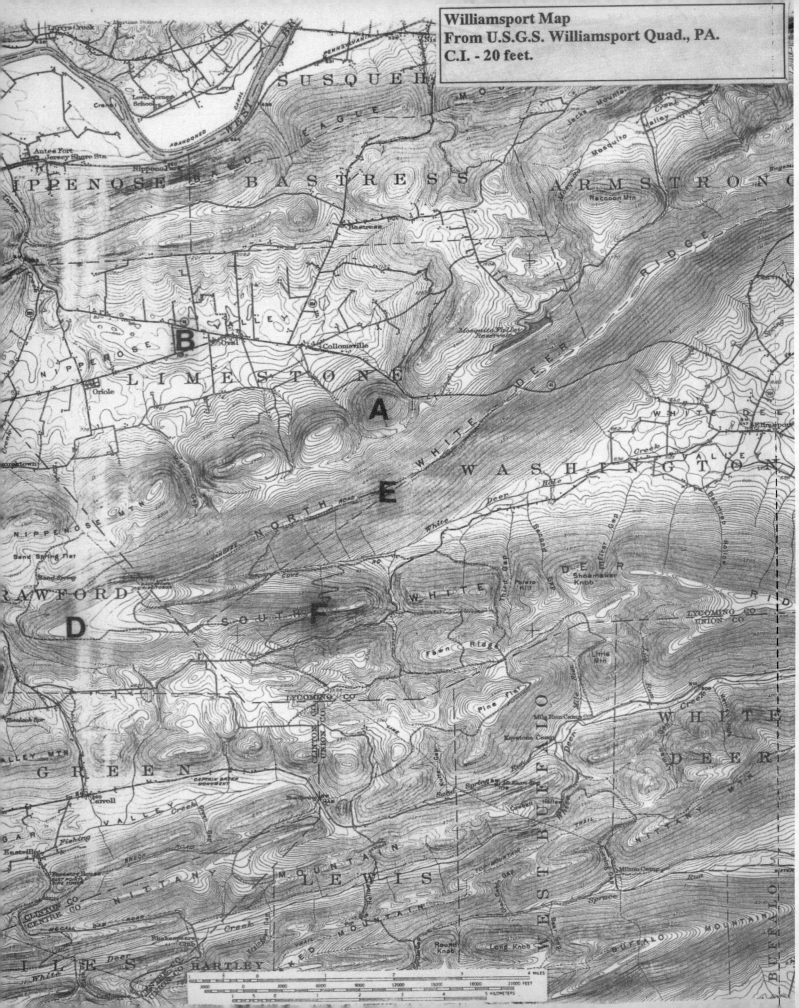

Milton Map
From U.S.G.S. Milton Quad., PA.
C.I. - 20 feet.

NAME _____

TOPOGRAPHIC EXPRESSION OF FOLDED STRATA – ANTELOPE RIDGE, WYOMING

1.   What diagnostic feature on Antelope Ridge (A) tells you its geologic structure?
_____
Antelope ridge is a(n) _____ (structure) plunging _____
Antelope Springs Creek cuts across Antelope Ridge at 'B'. What are two possible ways in which
the   creek   may   have   accomplished   this?   _____
_____

2.   What two topographic features along the ridge 'C' can be used to determine its geologic
structure? _____ _____ What is the geologic
structure there? _____ What is the *angle* of dip (in degrees) and the direction of
dip? _____ Show your calculations.

3.   What is the geologic structure (including direction of dip or plunge) at the following places:
At 'D'_____ The evidence for your answer is _____
At 'E' _____ The evidence for your answer is _____
_____

4.   Note the sloping surface in the vicinity of 'F'. Is this a cyclic or noncyclic surface?
_____ The evidence for your answer is _____
_____

5.   Note the flat-topped bench about 40' above Muddy Cr. at 'G'. Is this bench a cyclic erosion
surface or a stripped structural surface? _____ The evidence for your answer is
a. _____   b. _____

TOPOGRAPHIC EXPRESSION OF FOLDED STRATA –
ANTELOPE RIDGE, WYOMING

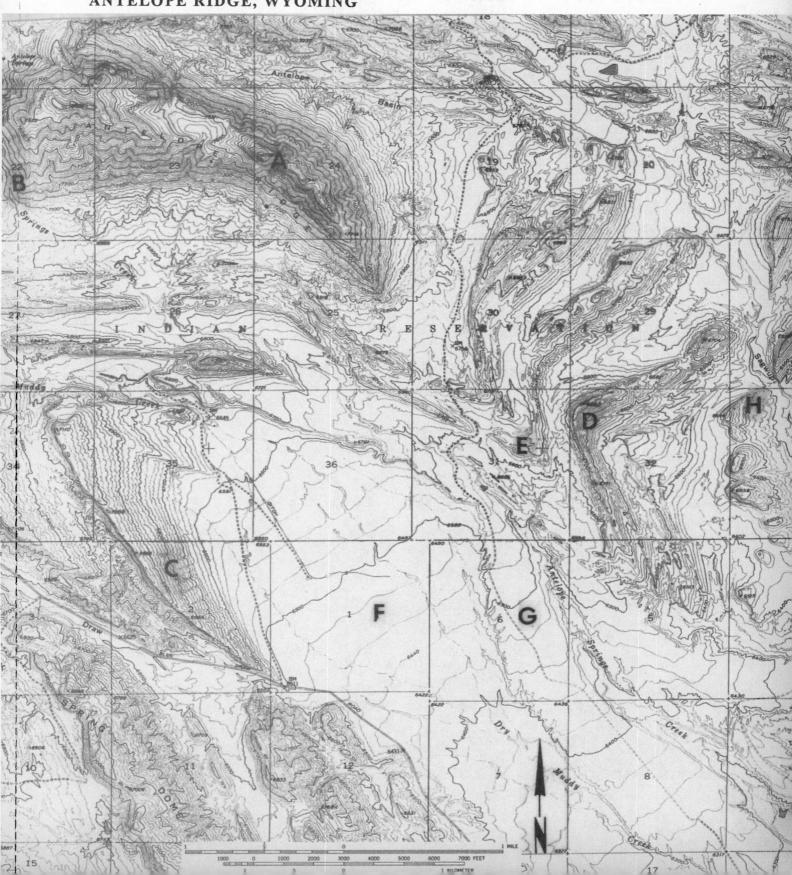

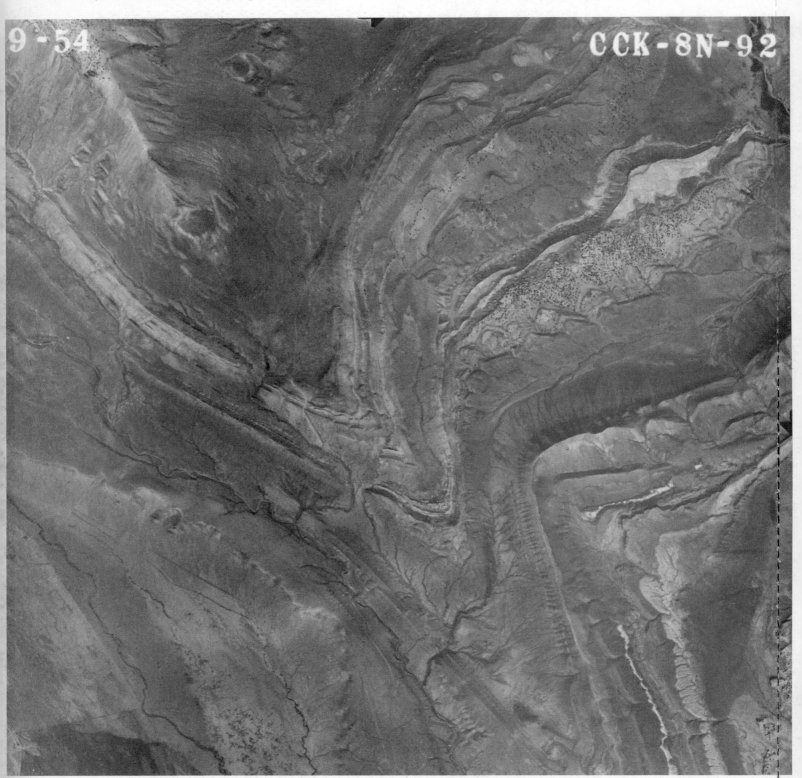

TOPOGRAPHIC EXPRESSION OF FOLDED STRATA –
ANTELOPE RIDGE, WYOMING

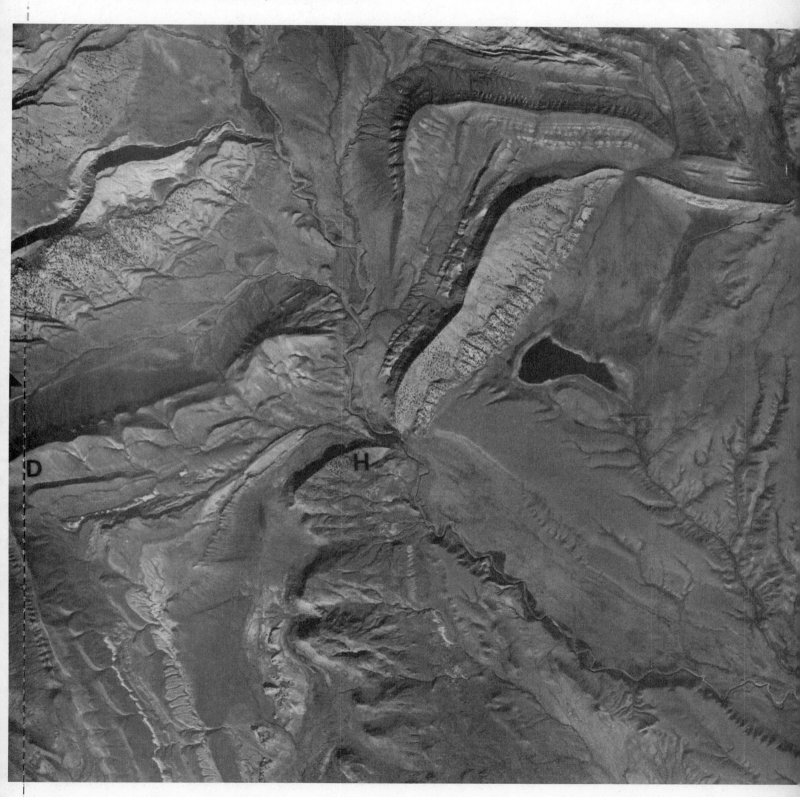

TOPOGRAPHIC EXPRESSION OF FOLDED STRATA – STRASBURG, VIRGINIA

1. What kind of rock makes up the ridges? _____ The valleys? _____
All of the ridges are continuously connected, so they must all be composed of the same bed.

2.   Check the ridges for asymmetry and V-shaped notches. Without good asymmetry and without V-shaped notches in the ridges, other evidence must be used to work out the geologic structure. The best place to look for diagnostic features is at the apex (nose) of plunging folds where the dip is usually gentle enough to show asymmetry. Note the crest of Little Crease Mt. (A). Does it have a gently-tapering nose or a steep, blunt nose? _____ Therefore, the geologic structure of Little Crease Mt. is _____ plunging _____ That means the valley of Mill Run (B) must be a _____ plunging _____. Check this by noting the nose of the fold at 'C'. Is it gently tapering or steep and blunt? _____

3.   Considering your analysis of the structure at Mill Run, the ridge at 'D' must then be a _____ (*structure*) _____ (*topographic form*) dipping _____

4. The bed making up the ridge at 'D' can be traced continuously around to the north where it becomes the ridge at 'E'. If the bed making the ridge at 'D' dips _____ and it is the same as the bed making the ridge at 'E', the geologic structure of the valley at 'F' must be a _____ Draw the axis of the structure on the map.

5.   Having worked out this part of the geologic structure of the area, now consider the geologic structure of the ridges at 'E' and 'G'. Start by noting the nose of the structure at Signal Knob (H). Is the nose gently tapering or steep and blunt? _____ The structure there must be a _____ plunging _____ Draw the axis of this structure on the map.   Thus, Three Top Mt. (G) must be a _____ (*structure*) _____ (*topographic form*) dipping _____, and Little Fort Valley (I) is a _____ (*structure*) _____ (*topographic form*).

6. This leaves only the structure of the ridge at 'E' to be determined. Although this can be done by continuing the same logic, an easy way to visualize the structure is by drawing a geologic cross-section along the line J-J', making sure that all of structures are continuous and consistent with your answers to the questions above. Use the profile below for your cross section.

J   J'

7. Mill Run flows part of the way in the valley at 'B', but instead of flowing out the lower end of this valley, the stream turns abruptly and cuts across Little Crease Mt. at Veach Gap. Can you offer an explanation for this anomalous behavior?

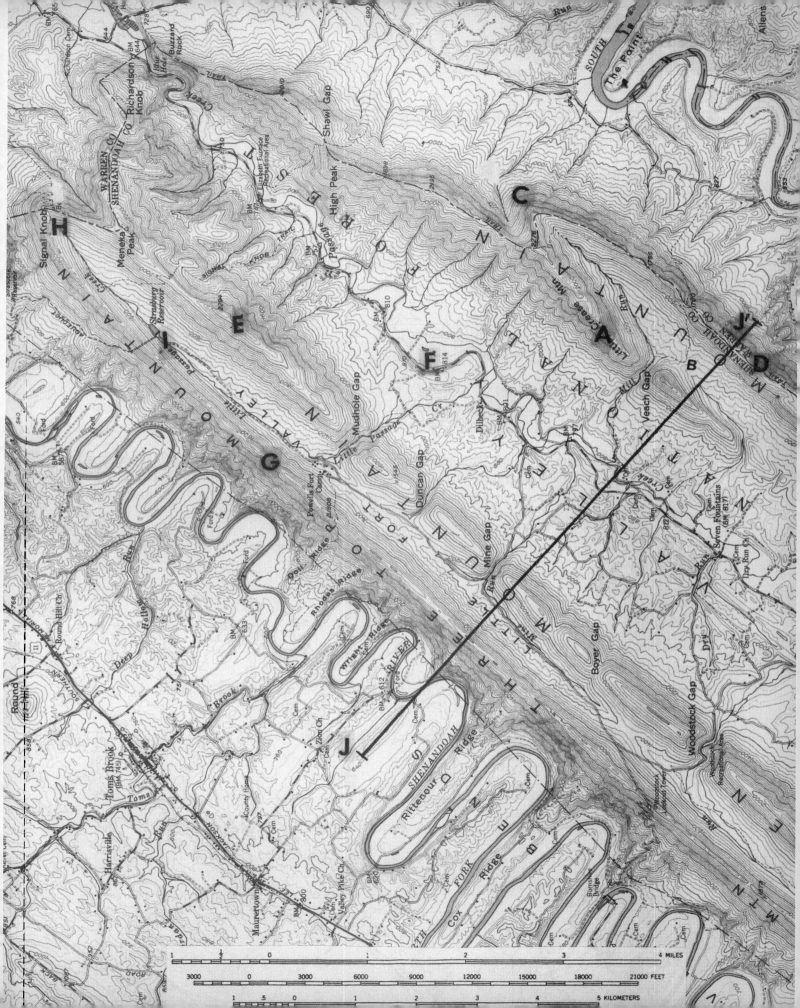

NAME _____

TOPOGRAPHIC EXPRESSION OF FOLDED STRATA – SPENCE, WYO.

1. What diagnostic topographic feature allows determination of the geologic structure of the ridge at 'A' ? _____ The geologic structure there is _____ dipping _____

2. What diagnostic topographic feature allows determination of the geologic structure of Sheep Mtn. (B)? _____ Sheep Mtn. is a _____ plunging _____ .

3. The bed at 'C' makes a _____ ridge that has been folded into a _____ (structure) plunging _____. What diagnostic topographic feature is evidence for your conclusion? _____

4. At 'D', the same bed as at 'C' has been folded into a _____ (structure) plunging _____. What diagnostic topographic feature is evidence for your conclusion? _____

5. Combining geologic structure with topographic form provides a concise way of expressing the relationship between topography and geologic structure (e.g., homoclinal ridge, homoclinal valley, anticlinal ridge, anticlinal valley, etc.). Name the following, using this combination of topographic form and geologic structure:

a. The valley at 'E' _____

b. The valley at 'F' _____

c. The valley at 'G' _____

d. The ridge at 'B' _____

e. The ridge at 'H' _____

6. What are two possible ways in which the Big Horn River may have established its course transverse to the structure at Sheep Mtn. (I)?

a. _____   _____

7. What are two possible origins of Red Flat (J)?

a. _____   b. _____

8. What topographic evidence on the map allows you to determine the origin of Red Flat?

a. _____   b. _____

9. What is the origin of Red Flat? _____

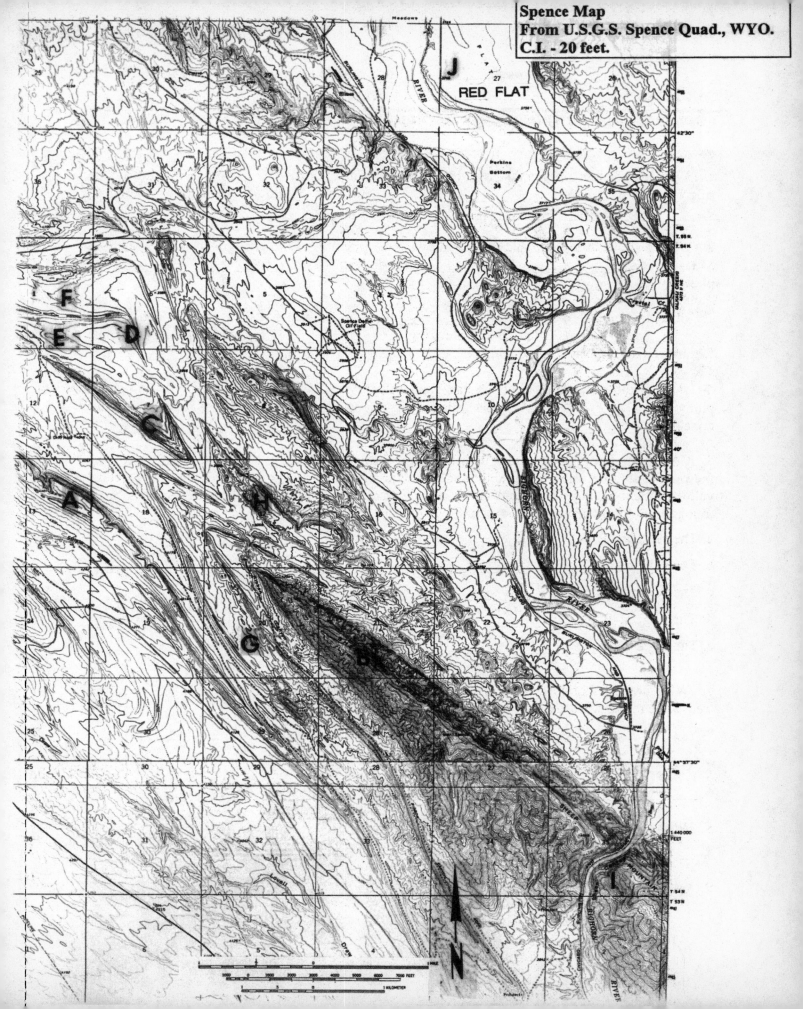

Spence Map
From U.S.G.S. Spence Quad., WYO.
C.I. - 20 feet.

RED FLAT

TOPOGRAPHIC EXPRESSION OF FOLDED STRATA – SPENCE, WYO.

NAME _____

TOPOGRAPHIC EXPRESSION OF FOLDED STRATA – HURRICANE, UTAH

1. The strike of the rocks at Harrisburg Bench (A) is _____ and the direction of dip is _____ based on _____ (evidence from map). This ridge is a _____ (structure) _____ (topography).

2. The strike of the rocks at 'B' is _____ and the direction of dip is _____ based on_____ and _____ (evidence from map). The ridge here is a _____ (structure) _____ (topography).

3. Considering your answers to questions (1) and (2), Purgatory Flat (C) is _____ (structure) _____ (topography), and Harrisburg Dome is _____ (structure) _____ (topography).

4. Warner Ridge (D) is a _____ (structure) _____ (topography), based on _____ and Warner Valley is a _____ (structure) _____ (topography).

5. On the air photo, note the beds NE of the ridge making Harrisburg Bench. Do they strike parallel to Harrisburg Bench? _____ Is Washington Black Ridge (E) also parallel to the strike? _____ What does this mean with regard to the structure of Washington Black Ridge? _____ Considering the sinuous nature of Washington Black Ridge and its flat, gently-sloping top, what is the most likely origin of it?
_____

6. What is the origin of Volcano Mtn.? _____ If the name were not so obvious, what evidence from the map or and air photo would lead you to the same conclusion?
_____

7. The ridge NE of Warner Valley (F) curves from N-S on the west side of the map to E-W at the center of the map, to N-S again near Sand Hollow Draw (G). This ridge is a _____ (structure) _____ (topography) that makes up a _____ (structure) plunging _____. The abrupt termination of the ridge against Hurricane Cliffs (H) and the straightness of the cliffs suggest that they have been formed by _____

8. Draw a geologic cross section along line I-I'.

I| |I'

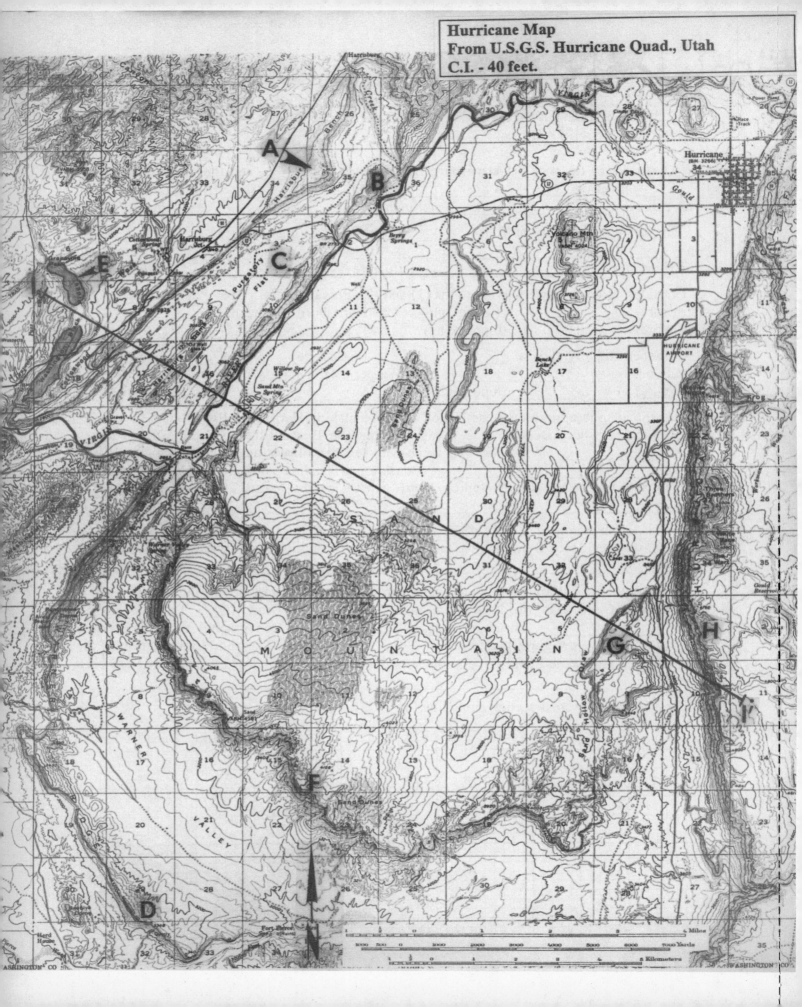

Hurricane Map
From U.S.G.S. Hurricane Quad., Utah
C.I. - 40 feet.

NAME _____

TOPOGRAPHIC EXPRESSION OF FOLDED STRATA – HURRICANE, UTAH

TOPOGRAPHIC EXPRESSION OF FOLDED STRATA –
HURRICANE, UTAH

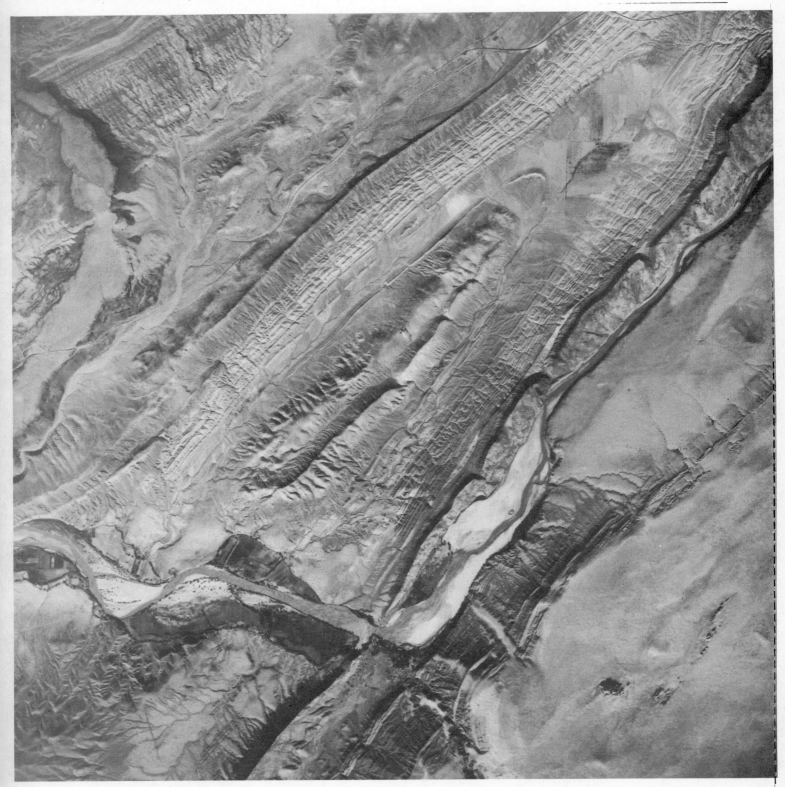

NAME _____

TOPOGRAPHIC EXPRESSION OF FOLDED STRATA – LOVELAND, COLORADO

1. The geologic structure of the ridge at 'A' is a _____ whose strike is _____ and the direction of dip is _____. Evidence for your answer
_____

2. The geologic structure of the ridge at 'B' is a _____ whose strike is _____ and the direction of dip is _____. Evidence for your answer
_____

3. The topography at 'C' is much more massive than at 'A' or 'B'. What does this suggest about the rock type at 'C'? _____

4. The pair of linear ridges at 'D' and 'E' are what you might expect of a breached anticline. However, are the dips of the ridges consistent with such an interpretation? _____
Why or why not? _____

5. Note the topographic expression of each pair of ridges, looking for distinctive features that distinguish it from the other ridges. Does the topographic expression of ridge 'A' more closely resemble that of ridge 'D' or 'E'? _____ Is this consistent with an anticlinal structure? _____

6. The profile below is drawn along line F-F'. Using the features you identified in the questions above, sketch in the only geologic structure that allows matching up of the ridges on opposite sides of the high area at 'C'.

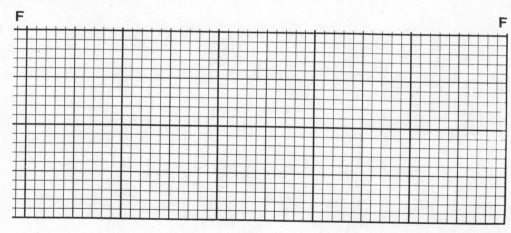

7. The zigzag bend that Ridge 'G' makes at Flatiron Mt. before continuing at 'H' lines up with a scarp in the massive topography at 'I'. What geologic structure is suggested by this?
_____

8. A similar scarp in the massive terrain occurs at 'J' (Green Ridge). If the strike of this scarp is extended, it intersects the abrupt termination of ridge 'A' at 'K'. What geologic structure is suggested by this? _____

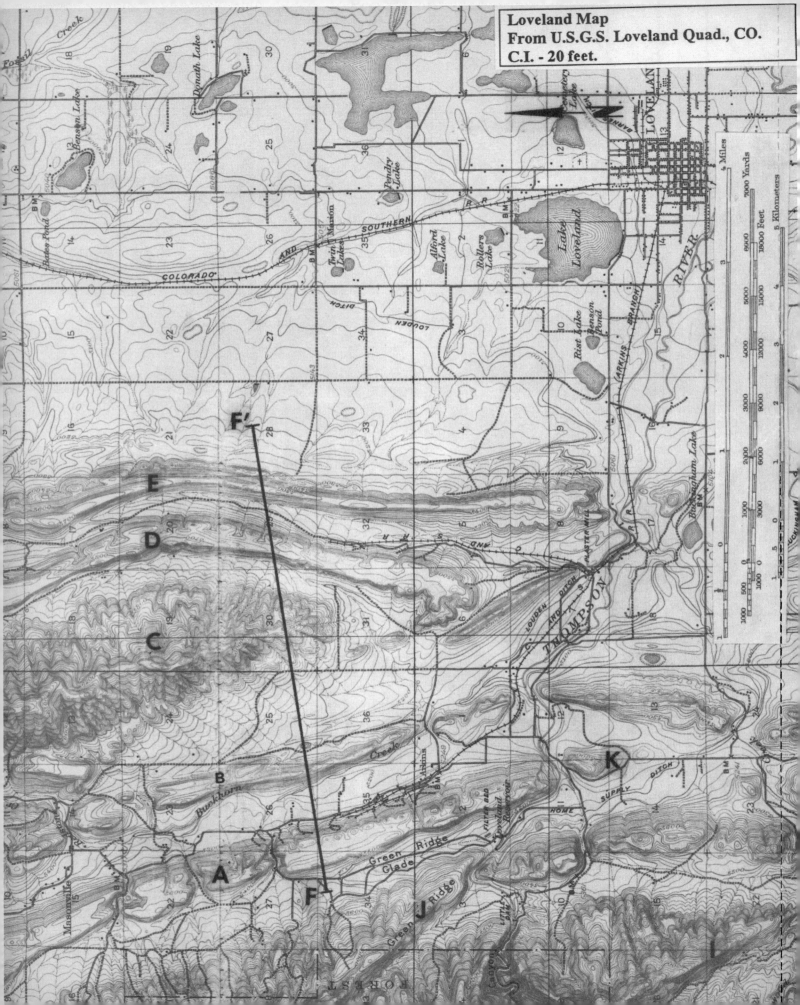

Loveland Map
From U.S.G.S. Loveland Quad., CO.
C.I. - 20 feet.

TOPOGRAPHIC EXPRESSION OF FOLDED STRATA –
LOVELAND, COLORADO

TOPOGRAPHIC EXPRESSION OF FOLDED STRATA –
LOVELAND, COLORADO

Topographic Expression of Joints and Faults

Criteria for Recognition of Faults

1. Scarps
    a. Fault scarps - offsetting of the land surface by faulting
    b. Fault-line scarps - differential erosion of rocks on opposite sides of a fault
        (1) Obsequent fault-line scarps
        (2) Resequent fault-line scarps
    c. Composite scarps

2. Triangular facets

3. Drainage changes
    a. Hanging valleys
    b. Wine glass structure
    c. Break in stream profiles (i.e., waterfalls, rapids)
    d. Ponding of streams against scarps
    e. Sag ponds
    f. Abandoned channels (changes in stream course)
    g. Alignment of stream courses; rectangular and angular drainage patterns
    h. Tilted terraces
    i. Selective headward erosion along fault zones
    j. Offsetting of streams
    k. Shutter ridges

4. Alluvial fans
    a. Fresh fans
    b. Scarps in fans

5. Springs

6. Offset strata

7. Fault drag

8. Truncation of strata; resistant ridges; discontinuity of structures

TOPOGRAPHIC EXPRESSION OF JOINTS AND FAULTS – JELLICO, CALIFORNIA

1. Trace the trend of the scarp face at Creek Rim (A) along its extent. Does its height remain constant? _____ Does the scarp face remain as a single cliff on the right hand map? _____ What happens to it? _____ Are these features consistent with the topographic expression of a single resistant sedimentary bed? _____ What geologic structure does explain the topographic expression of Hat Creek Rim?
_____

2. What is the origin of the scarp at Butte Creek Rim (B)? _____
What evidence can you cite for your conclusion?
(a.) _____ (b.) _____

3. What is the origin of the scarp at 'C'? _____
What evidence supports your conclusion? _____

4. What is the origin of the scarp at 'D'? _____
What evidence supports your conclusion? _____

5. Consider the possibility that the scarps are fault scarps. If that were true, what would be the vertical displacement at the following places?
a. At 'E' _____
b. At 'F' _____
c. At 'G' _____
d. At 'H' _____
e. At 'A' _____
Are these variable heights consistent with a fault-scarp origin? _____

6. What does the stipple pattern at (I) represent? _____

7. What controlled the northernmost tongue (I)? _____

8.   What does this suggest regarding the age of the tongues relative to faulting in the area? _____ Why? _____

9. Does this relationship hold for the lava at Hat Creek Rim (K)? _____

10. What are the conical-shaped hills (J) in the stippled area? _____

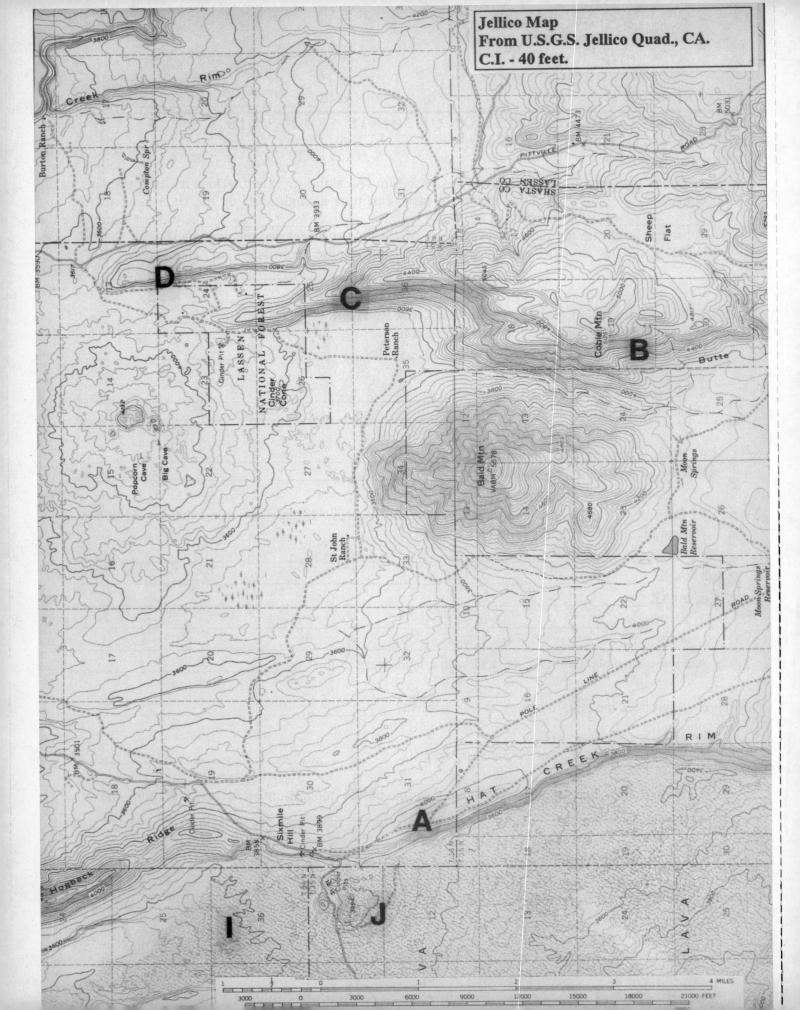

Jellico Map
From U.S.G.S. Jellico Quad., CA.
C.I. - 40 feet.

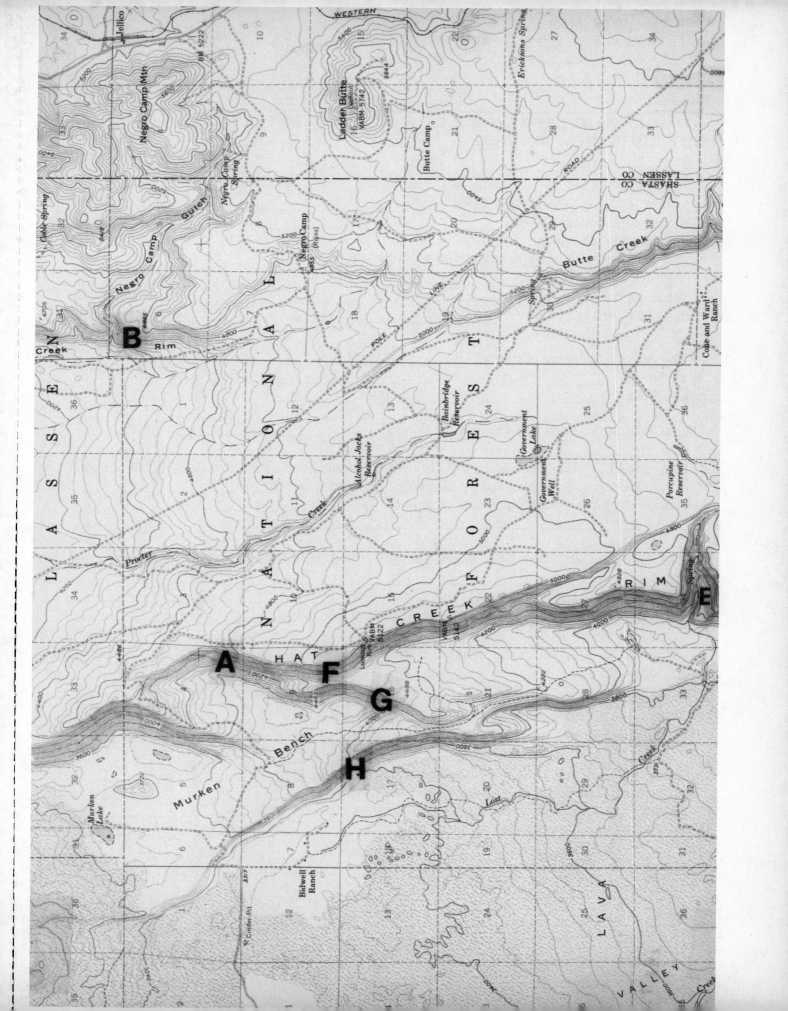

EXERCISE 10.1   NAME _____

**TOPOGRAPHIC EXPRESSION OF JOINTS AND FAULTS –
SHIP ROCK, NEW MEXICO**

1. What is the origin of Ship Rock (A)? _____   What topographic
evidence supports your answer? _____
How might you tell whether Ship Rock is the result of the accumulation of material on the earth's
surface or is the result of differential erosion of resistant rock? _____

2. What are the linear ridges (B) near Ship Rock? _____   What topographic
evidence supports your answer? _____ _____

3. The geologic structure of ridge 'C' is a _____ dipping _____
 as shown by _____ and _____ The geologic structure of
ridge 'D' is a _____ dipping _____ as shown by
_____ _____
The geologic structure of Little Ship Rock Wash (E) is a _____ dipping _____

4. Mitten Rock (F) is an anomalous mass of rock in the middle of Little Ship Rock Wash. Look at
it on both the topographic map and the air photo. How did it form? _____
What topographic evidence supports your answer? _____
_____

5. The double ridges (C and D) terminate abruptly at Red Rock Highway (G). What is the reason
for this? _____
On the opposite side of the highway, ridge 'C' becomes _____
On the opposite side of the highway, ridge 'D' becomes _____

6. What is the direction of dip of the bed making Rock Ridge at 'H' _____. Thus, the
geologic structure at 'I' is a _____ plunging _____

7. What is the direction of dip of the beds making Rock Ridge at 'H'? _____
Thus, the geologic structure there is a _____ plunging _____

8. What is the geologic structure of the Beautiful Mts.? _____
What topographic evidence supports your answer? _____
Is this structure consistent with the geologic structure to the area to the north? _____
What are two possible geologic structures here? _____
_____

SHIP ROCK, NEW MEXICO

9. Note the cliffs near Red Wash (K)? What is the direction of dip here? _____
Comparing this to the beds to Shiprock Wash, what happens to the dip? _____

10. What kind of rock makes up the broad plain between 'A' and 'C'? _____
Calculate the dip of the beds making the ridge at 'C' _____
If this dip extends all the way to Shiprock, how thick would the rock unit there have to be?
_____. Is this a reasonable thickness for a rock unit of uniform composition? _____
What does this tell you about the dip of the beds making up the eastern half of the map?
_____
_____

11. Draw a geologic cross section along the line L-L'.

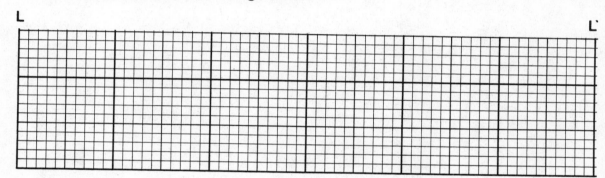

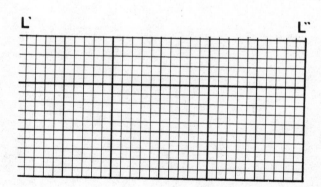

12. A stream appears on the surface at 'M', flows on the surface to 'N', then disappears again.
Note the stipple pattern at the word "Wash" (near 'M'). Where does the water come from that
suddenly appears at 'M'? _____ Why does it flow on the surface for only a
short distance? _____

89

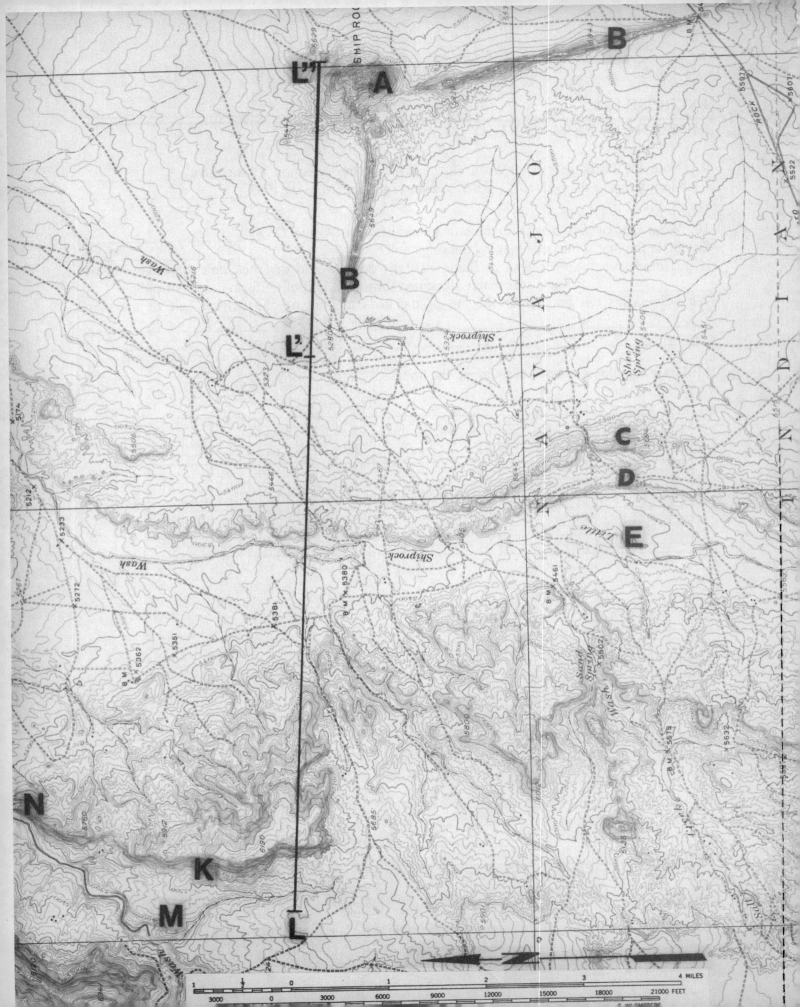

NAVAJO

INDIAN

SHIP ROCK

Shiprock

Sheep
Spring

Shiprock

Sand
Spring
Wash

Wash

Wash

Wash

A

B

B

L"

L'

C

D

E

N

K

M

L

4 MILES

3000   0   3000   6000   9000   12000   15000   18000   21000 FEET

1   ½   0   1   2   3   4 MILES

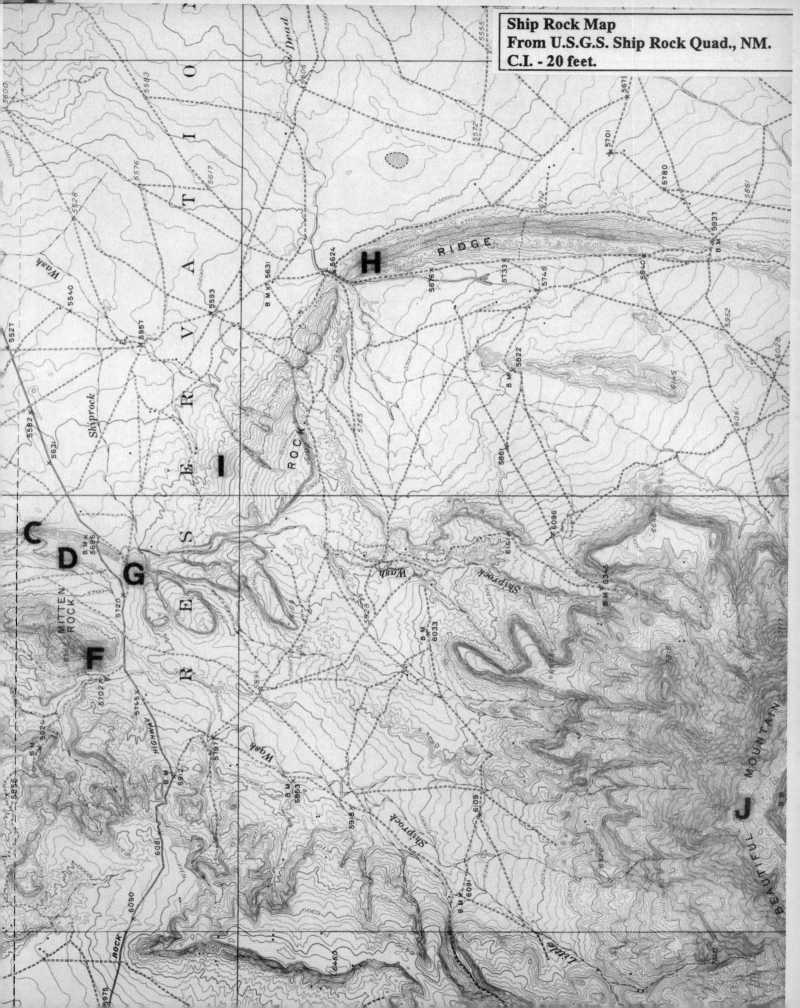

TOPOGRAPHIC EXPRESSION OF JOINTS AND FAULTS – SHIP ROCK, NEW MEXICO

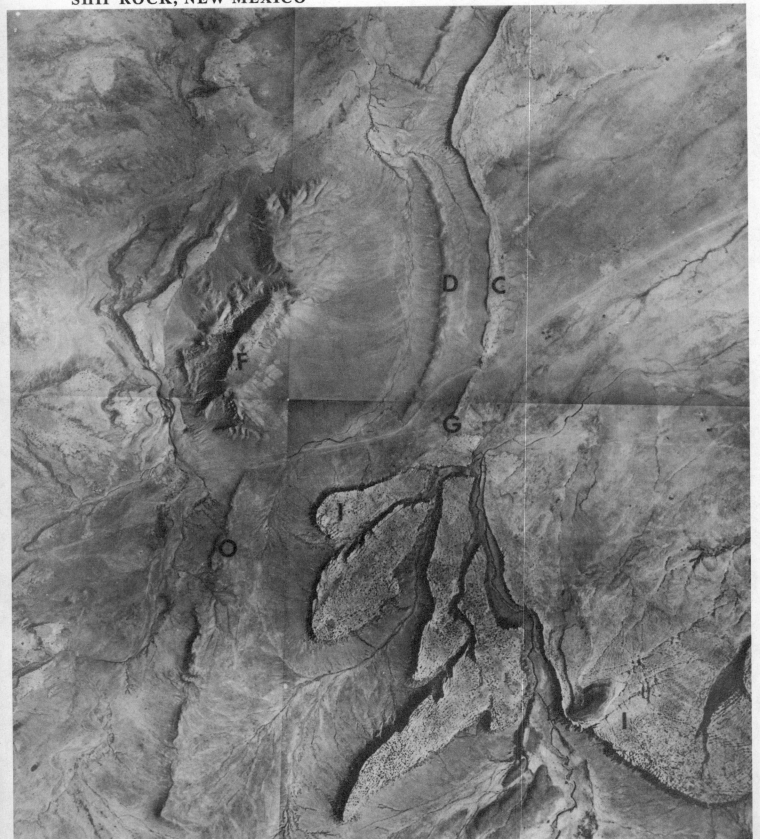

TOPOGRAPHIC EXPRESSION OF JOINTS AND FAULTS – SHIP ROCK, NEW MEXICO

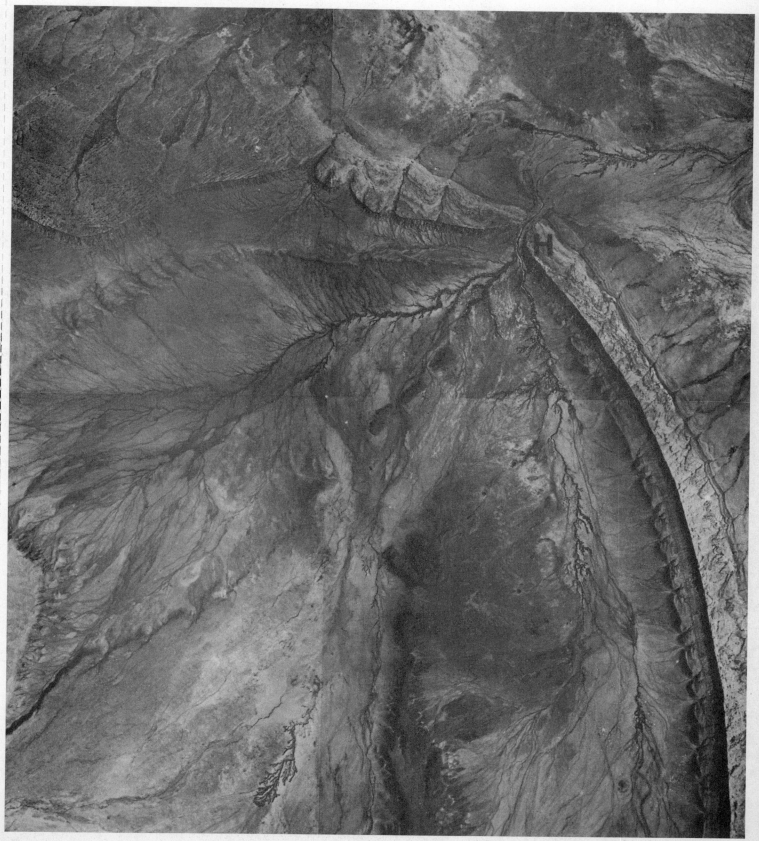

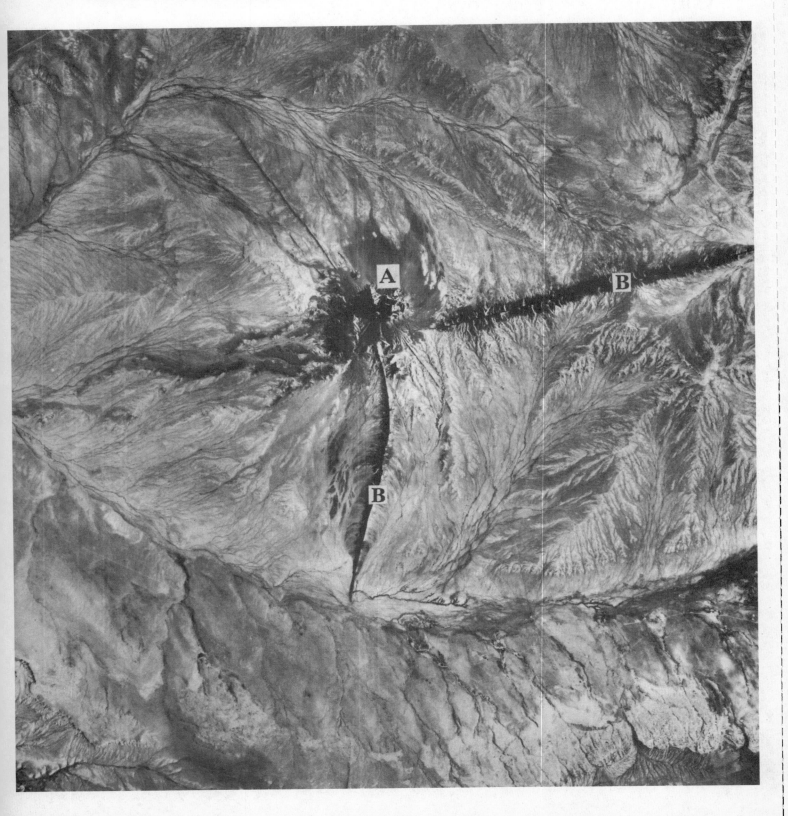

NAME _____

TOPOGRAPHIC EXPRESSION OF JOINTS AND FAULTS –
MT. DOME AND TULELAKE, CALIFORNIA

1. Note the east-facing scarps at 'A', 'B', 'C', and 'D'. West of each scarp are more gently sloping surfaces that, together with the scarps, resemble several parallel homoclinal ridges. However, westward-flowing streams at the base of scarp 'B' have been dammed by the scarp to form Gillem Lakes and westward-flowing streams at the base of scarp 'C' have been dammed by the scarp to form Crumes Lake. Is this consistent with interpretation of the scarps as homoclinal ridges? _____ Why or why not? _____

2. Considering your answer to question 1, what must the scarps represent? _____
Why? _____

3. Note that scarp 'C' bends to the NW and decreases in height until it disappears altogether at 'F', and another scarp at 'E' continues on strike with scarp 'C'' before also bending to the NW. Two other scarps, 'E' and 'G' parallel the scarp at 'F'. Is this consistent with your answer to question 2? _____

4. Considering the dammed drainages, could these be fault-line scarps? _____
Why or why not? _____

5. If the scarps are a result of direct offsetting of the land surface, what is the maximum amount of displacement on each fault? 'A' _____ 'B' _____ 'C' _____ 'D' _____

6. Are the lava flows (H) at the base of Gillem Bluffs older or younger than the scarp? _____
How can you tell? _____

7. What is responsible for the flat, very low-relief plain at 'I'? _____
If you drilled a well into it, what would you expect to find? _____

8. What is the origin of the many west-facing scarps on the east side of the Tulelake map?
_____ What evidence from the map supports your answer?
(a.) _____ (b) _____

9. What does the direction that the scarps on the Mt. Dome map face and the direction that the scarps on the Tulelake map face, suggest about the geologic structure beneath the Tulelake plain (I)? _____

10. What does the shape of Horse Mtn. (J) suggest about its origin? _____

Mt. Dome Map
From U.S.G.S. Mt. Dome Quad., CA.-OR.
C.I. - 40 feet.

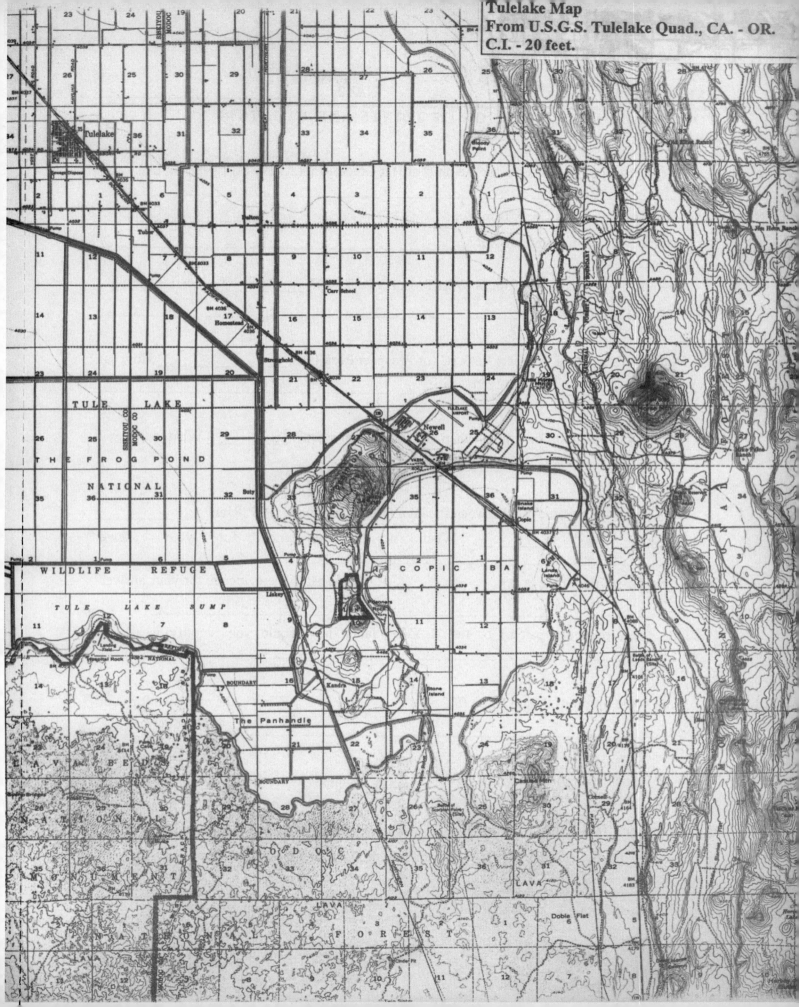

Tulelake Map
From U.S.G.S. Tulelake Quad., CA. - OR.
C.I. - 20 feet.

NAME _____

TOPOGRAPHIC EXPRESSION OF JOINTS AND FAULTS – PROSPECT PEAK, CALIFORNIA

1. What are two *possible* origins for the bluff at Hat Creek Rim (A)?

a. _____ b._____

What *is* the origin of Hat Creek Rim? _____

What is the evidence for your answer? _____

2. Which is older, Hat Creek Rim or West Prospect Peak (B)? _____ What is the

evidence for your answer? _____

3. Which is older, West Prospect Peak (C) or Prospect Peak (D)? _____ How can

you tell? _____

4. Is Prospect Peak older or younger than Hat Creek Rim scarp? _____ What is the

evidence for your answer? _____

5. What is the age relationship between the lava flow at (E) and Hat Cr. Rim? _____

_____ What is the evidence for your answer?

_____

What was the source of the lava? _____ What is the evidence for your answer? _____

_____

Without the brown dot pattern or the "LAVA" label, how could you identify it as lava?

_____ _____ _____

6. What is the origin of Potato Butte (F)? _____ What is the evidence for your

answer? _____

7. What is the origin of Badger Mt. (G)? _____ What are the linear scarps on it? _____

8. What is the origin of Hat Mt. (H)? _____ What caused the gap in its east side?

_____ How did the topography for ~2 miles to the east form?

_____

9. What caused Song Lake (I) to form? _____ Evidence for your answer? _____

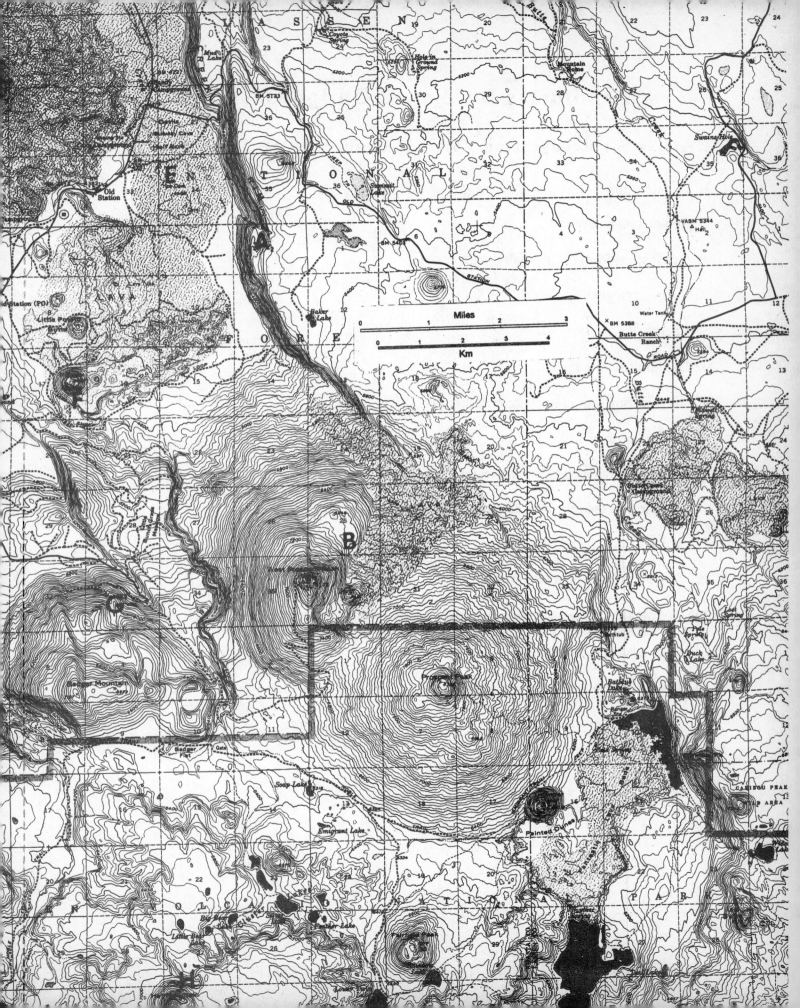

Landforms Developed On Igneous Rocks

Criteria for Recognizing Volcanic Landforms

1. Lobate form

2. Sinuous map pattern

3. Originating from or associated with cones

4. Pressure ridges

5. Levees

6. Digitate margins

7. Irregular surface; many jagged contours

8. Lack of surface drainage

9. Disrupted drainage; ponded streams

10. Linear depressions from collapsed lava tubes

11. Rifts

12. "Twin" streams in a single valley

13. Topographic inversion (sinuous ridges capped with resistant lava)

14. Dark photographic tones

15. Unconformable across underlying rocks

16. Flows abut (or flow around) pre-existing topographic features

17. Glass flows have steep flow margins

EXERCISE 11.0   NAME _____

LANDFORMS DEVELOPED ON IGNEOUS ROCKS – JAPAN

1. Using the letters below, mark on the photos examples of the following:
   A. Lobate form   B. Source at cone   C. Pressure ridges
   D. Levees   E. Digitate margins   F. Irregular surface
   G. Linear depressions

2. Rank the relative age of flows; a, b, c. (oldest) _____ _____ _____ (youngest).
   How can you tell?_____

101

NAME _____

LANDFORMS DEVELOPED ON IGNEOUS ROCKS –
CRATERS OF THE MOON NATIONAL MONUMENT, IDAHO

1. What is the origin of the conical hills with concentric contours? _____

   What is the evidence for your answer? _____

2. What happened to missing half of Half Cone (A)? _____

3. Why are these hills lined up in a NW-SE direction? _____

4. Why are the contours at 'B' so jagged and crenulated? _____

5. What was the direction of flow at 'B'? _____

6. How did the linear features at 'C' form? _____

7. What is the Great Rift (D)? _____

8. Is the Great Rift older or younger than Fissure Butte (D)? _____

   How can you tell? _____

9. How did Dewdrop Cave and Indian Tunnell (E) form? _____

10. How did Big Sink Waterhole (F) form? _____

11. Tree molds occur at several places on the map (G). How do they form? _____

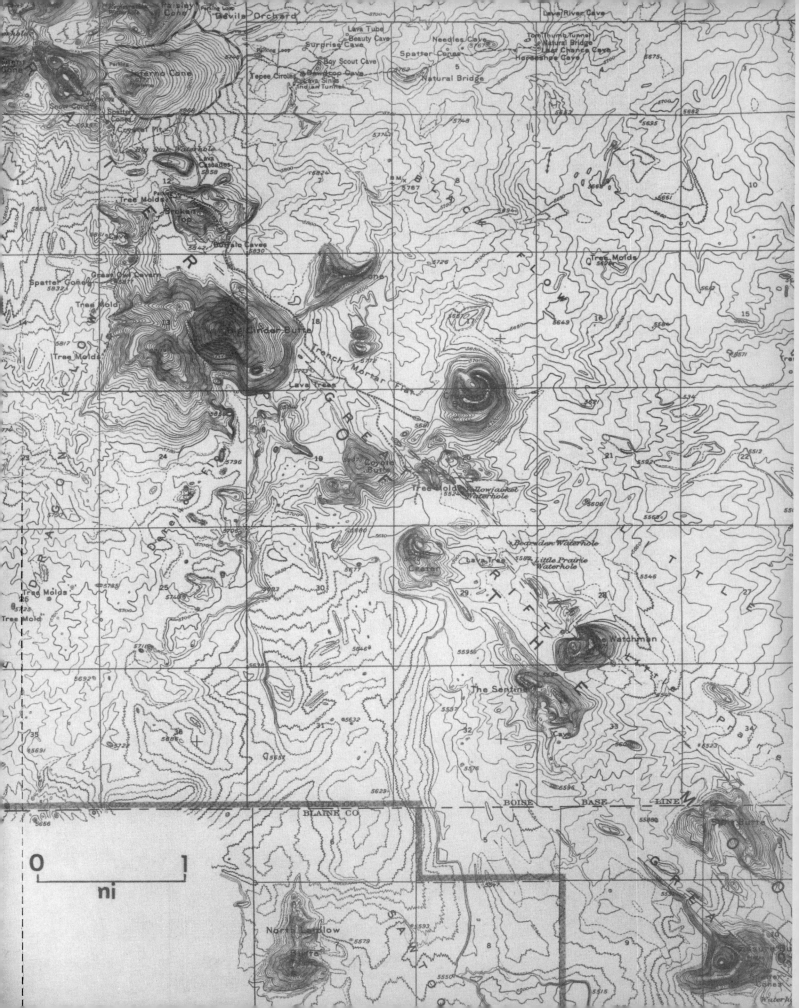

NAME _____

LANDFORMS DEVELOPED ON IGNEOUS ROCKS –
CRATERS OF THE MOON NATIONAL MONUMENT, IDAHO

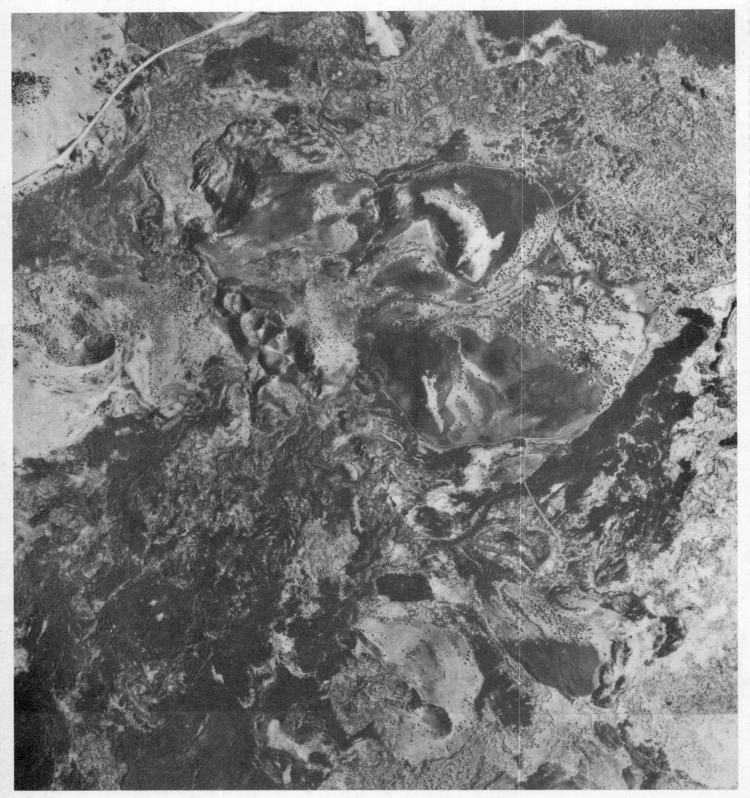

NAME _____

LANDFORMS DEVELOPED ON IGNEOUS ROCKS – MEDICINE LAKE AND TIMBER MOUNTAIN, CALIFORNIA

1. Does the lava flow at 'A' emanate from a typical volcanic cone? _____ Describe the form of the source of the flow at 'B' _____

What name may be given to such a feature? _____

2. From the height of the flow margins, what can you say about the viscosity of the lava? _____ What factors control the viscosity of a lava flow?

(a) _____ (b) _____ (c) _____

Compare the topographic expression of the flow at 'A' with those on the Craters of the Moon photo. Which of these is most silicic (dacite/rhyolite) _____ Which is basaltic? _____

3. What are the dome-like features at 'C'? _____

4. Which of the criteria for identification of lava flows can you identify on this photo?

(a) _____ (b) _____ (c) _____

(d) _____ (e) _____

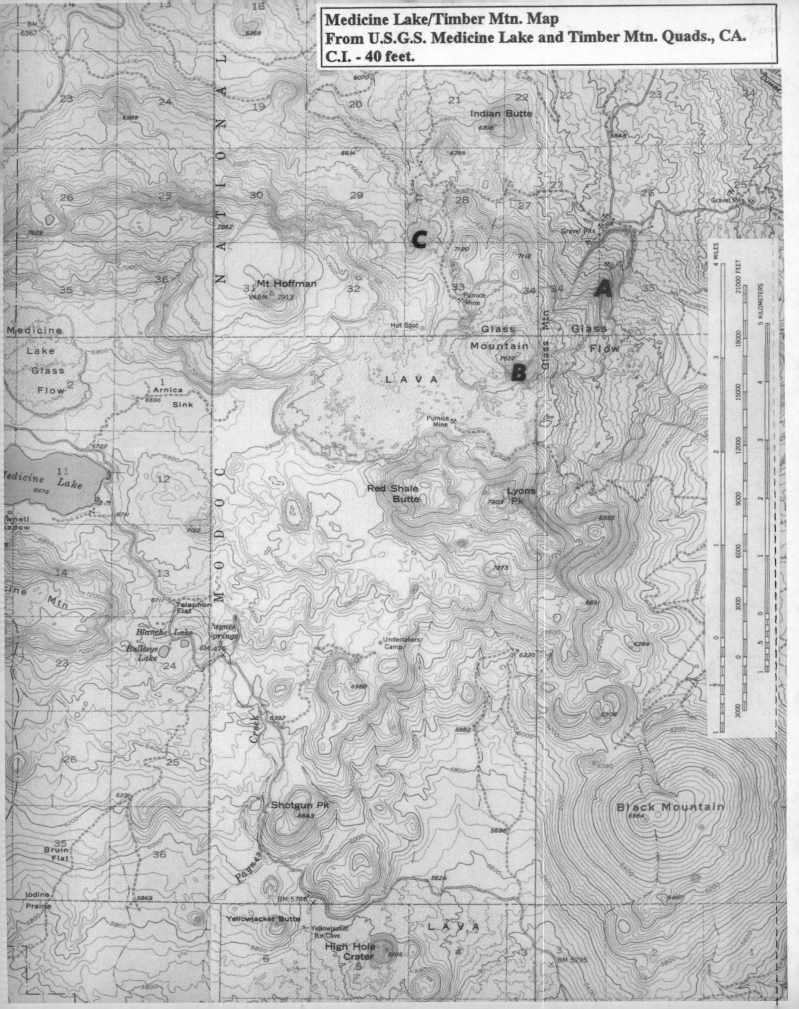

Medicine Lake/Timber Mtn. Map
From U.S.G.S. Medicine Lake and Timber Mtn. Quads., CA.
C.I. - 40 feet.

NAME _____

LANDFORMS DEVELOPED ON IGNEOUS ROCKS –
MEDICINE LAKE AND TIMBER MOUNTAIN, CALIFORNIA

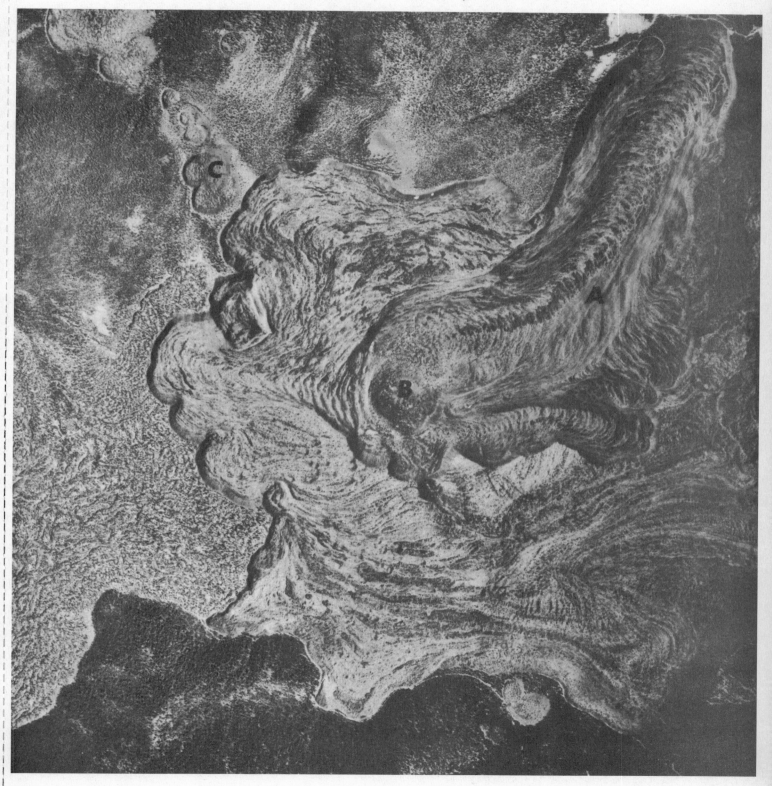

NAME _____

LANDFORMS DEVELOPED ON IGNEOUS ROCKS –
MAUNA LOA NATIONAL PARK, HAWAII

1. Mokuaweowea Crater (A) is _____ mi. long and _____ mi. wide and is an example
of a _____

2. In general, what are the most common causes for such large craters? (a) _____
(b) _____ Knowing that lavas in Hawaii are dominantly basaltic, which of the
possible causes of the large crater is most likely here? _____

3. Note the cusp-shaped margin of the large crater at 'B' and the smaller craters are 'C'. What
does this suggest about the origin of Mokuaweowea Crater? _____

4. What do the dashed lines at 'D' represent? _____

5. Why are the contours (E) so crenulated? _____

6. Mauna Loa is a classic example of a shield volcano. What is its angle of slope? _____
What does most of the cone consist of? _____

7. What do the shapes of the lava flows of 1851 (F) and 1926 (G) suggest about the way in
which the cone of Mauna Loa has been built? (e.g., did the lava flows erupt as sheets?)
_____

**Mauna Loa National Park Map
From U.S.G.S. Mauna Loa Quad., HW.
C.I. - 50 feet.**

GLACIAL PROCESSES AND LANDFORMS

Recognizing Glacial Features

The exercises that follow are based on air photos that portray features of modern glaciers. The object of these exercises is to recognize the types of landforms developed by various glacial processes.

The state of mass balance of a glacier (positive, negative, or neutral) may be determined by physical characteristics of the glacier, as listed below.

Active or Neutral	Negative
Bold, steep, well-defined terminus	Poorly defined terminus, commonly difficult to identify
Clean ice at terminus	Terminus covered with rock debris
Smooth glacier surface	Rough, irregular surface
Surface lakes absent	Circular, ice-walled lakes on the surface of the glacier

Because glaciers move dynamically, they commonly show surface features that reflect the movement of ice beneath. Among such features are crevasses and ogives.

Glacial Landforms

Erosional Landforms	Depositional Landforms
Cirques	Moraines
Aretes	Drumlins, flutes
Horns	Eskers
Glacial troughs, fjords	Kames
Cyclopian stairs	Ice-walled lake plains
Hanging valleys	Disintegration ridges and trenches
Finger lakes	Outwash plains

GLACIAL PROCESSES – ALASKA

1. The "bite" out of the glacier terminus at 'A' is caused by _____

2. Why is the other side of the glacier terminus at 'B' different? _____

3. What type of crevasse pattern occurs at 'C'? _____

4. Note the crevasse pattern between 'A' and 'C'. Why is it different than up glacier from 'B'?
_____

5. What kind of crevasse pattern occurs at 'D'? _____

6. Note what happens to the ice of the two tributary glaciers at 'E' and 'F' where they join. Name the feature at 'I' _____ Which tributary has more vigorously moving ice? _____
What happens to the ice from tributary 'F' as it approaches the terminus? _____
What is the reason for this? _____

7. What are the transverse, convex-down valley, dark bands in the middle of the small glacier at 'G'? _____ What kind of crevasse pattern occurs at 'H'?
_____

8. What is responsible for the sharp trim line at 'J'? _____
What does it indicate about the former extent of the glacier downvalley? _____
_____

GLACIAL PROCESSES – ALASKA

GLACIAL PROCESSES – ALASKA (14)

1. Locate the terminus of the glacier and mark it on the photo.

2. What makes the glacier so dark? _____

3. Why does the large lake at 'A' have cusp-shaped margins? _____

4. Why is the glacier surface so uneven? _____

5. Does the meltwater stream at 'B' emerge from the ice terminus? _____ Where does it originate (mark it on the photo)? _____ What does this indicate about the rate of ice movement? _____

6. The features above indicate that the mass balance of the glacier is _____

EXERCISE 12.2

ALASKA (15)

1. What are the dark, winding bands on the glacier surface on photo Alaska 15, 1954?

    _____

    Why are they so distorted? _____

2. Locate the glacier terminus on ALASKA-15-1949 and mark it on the photo. Judging from the appearance of the terminus, was the mass balance of this glacier positive, neutral, or negative in 1949? _____ What features on the photos did you base this conclusion on?
(a) _____ (b) _____ (c) _____

3. Note the location of several prominent streams on the valley side at 'A' and at 'B' on both photos. Using these streams for reference, note the distance of point 'C' on the glacier in 1949 relative to point 'A' and find the same point (C) on the 1954 photo.

a. What difference do you see between 1949 and 1954 on this glacier? _____

b. Note the ice-walled lake on the glacier surface at 'D' on the 1949 photo. Has it moved in the 1954 photo? _____ How can point 'C' have moved down valley when point D hasn't?

    _____

c. A number of ice-walled lakes occur on the glacier surface at 'E' on the 1949 photo. Where are they in 1954? _____

d. Where is the *active* glacier terminus in 1954? _____

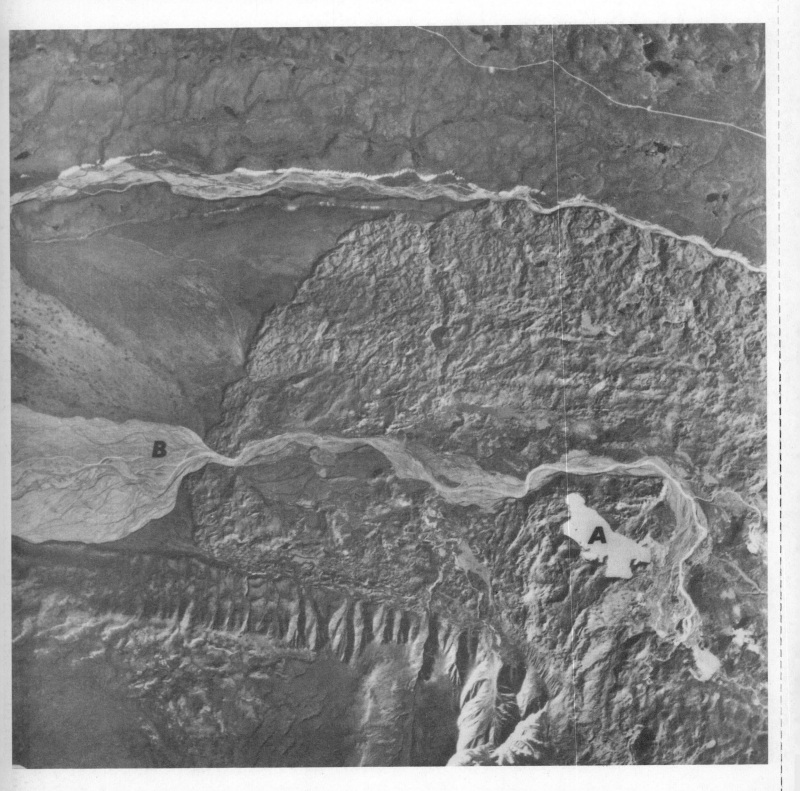

GLACIAL PROCESSES – ALASKA (15)

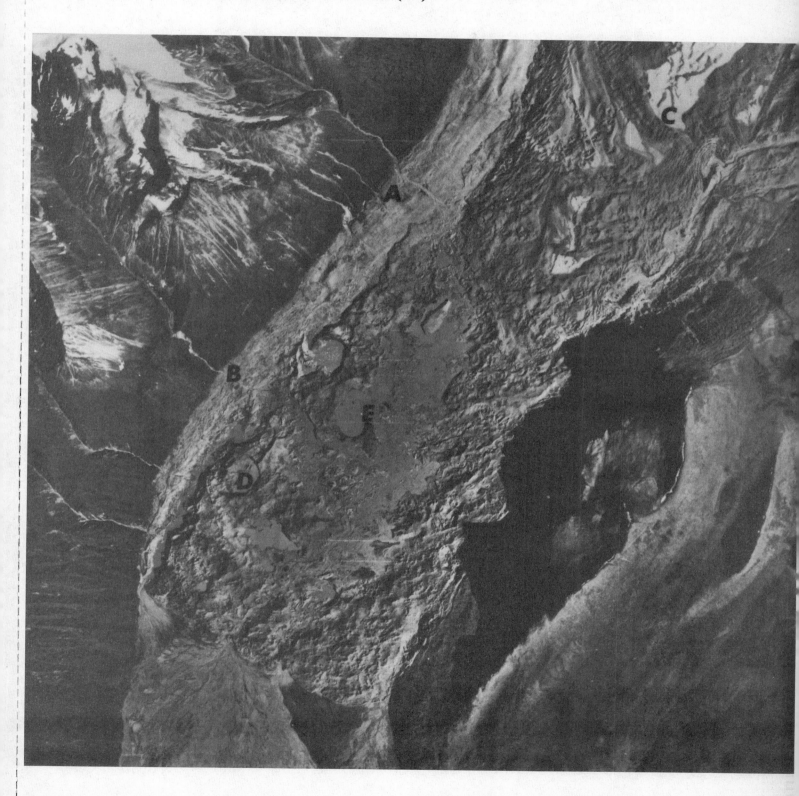

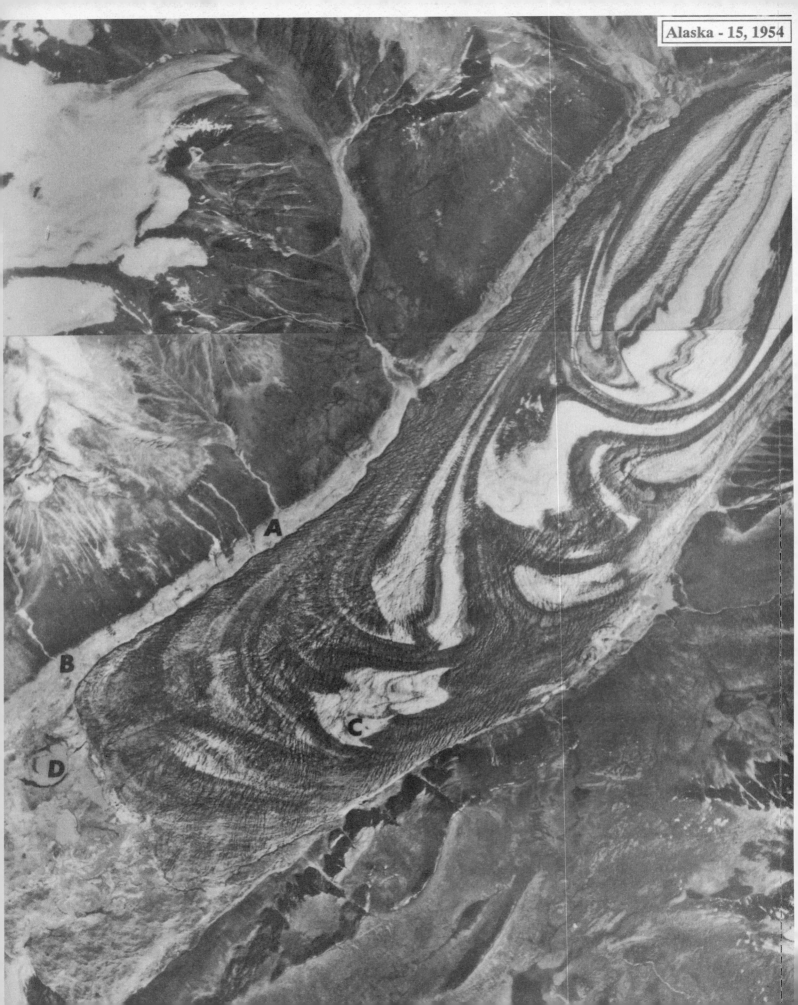

Alaska - 15, 1954

GLACIAL PROCESSES – ALASKA

Two glaciers are shown on the photograph; one with a broad terminus and one with a long, thin one.

1. Three distinct zones are evident near the terminus of the broad glacier. An inner zone (A) with several large lakes. Note that the large lakes have cusp-shaped margins and many cusp-shaped scarps occur to the right of the lakes. These cusp-shaped features are formed by
_____

An intermediate zone (B) with streams crossing it. An outer zone (C) with dark tone. This zone is darker than the others because of the presence of _____
What is the origin of the small lakes in this zone? _____

Note the large, cusp-shaped scarps (D) in the northernmost of the two southern, dark ridges in this zone. They are remnants of _____

2. Locate the terminus of the glacier. Mark it on the photo. What criteria did you use to find it?
_____

3. List the features on this glacier that indicate its state of mass balance.
a. _____ b. _____ c. _____

For the long, thin glacier (E):

4. Why is the surface so irregular? _____

5. List the features on this glacier that indicate its state of mass balance.
a. _____ b. _____ c. _____

6. Note how far up valley these features extend. What significance does this have with regard to the mass balance of the glacier? _____

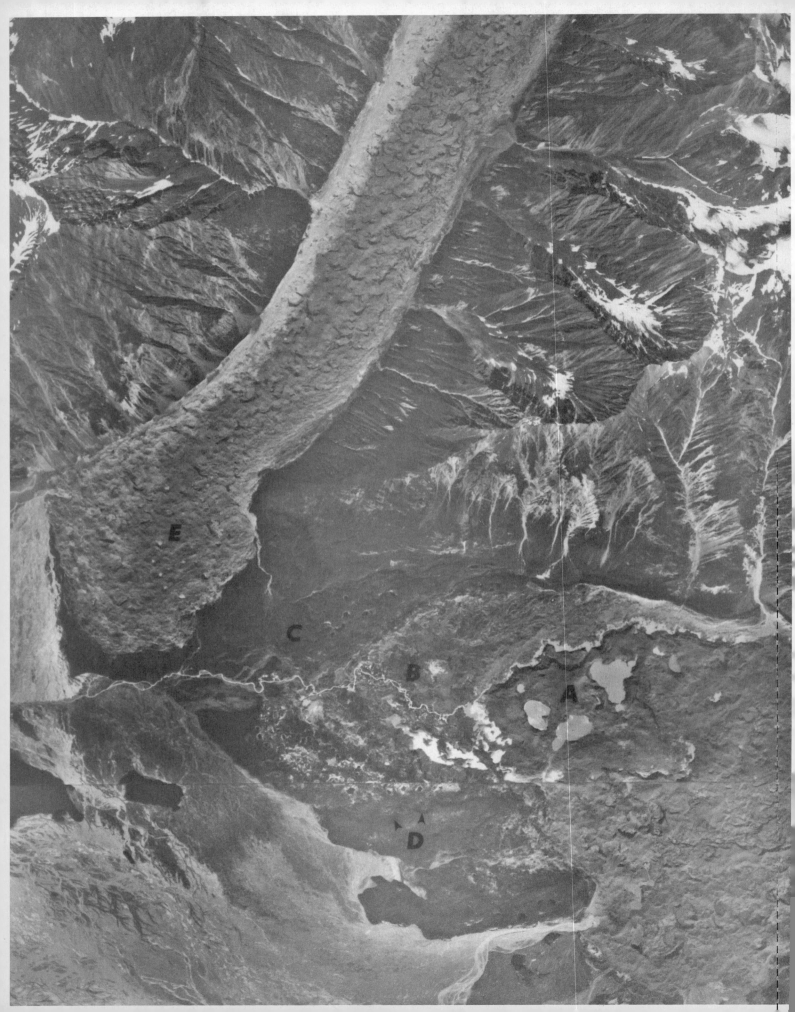

GLACIAL LANDFORMS – ALASKA

Most of the upper part of the photo is a glacier. The white area on the left is glacial ice and the white circular features at 'A' are ice-walled lakes.

1. Draw a line marking the terminus of the glacier on the photo. What features do you use to identify the terminus? (a) _____ (b) _____ (c) _____
Is the terminus well defined? _____ Why or why not? _____
_____

2. What is the dark area at 'B' _____ Is it part of the glacier? _____
How can you tell? _____
_____

3. Is the mass balance of this glacier positive, neutral, or negative? _____
List the evidence for your conclusion. (a) _____
(b) _____ (c) _____
_____

4. Does the channel at 'C' presently have water flowing in it? _____ What has happened to the channel? _____
Name the landform of which the channel is a part _____

5. Name the landform at 'D' _____ Is it actively forming now? _____
Is the landform at 'E' similar? _____ Is it actively forming now? _____
What has happened to it? _____

6. What does the linear topography at "F" suggest about the former activity of ice in the area?
_____

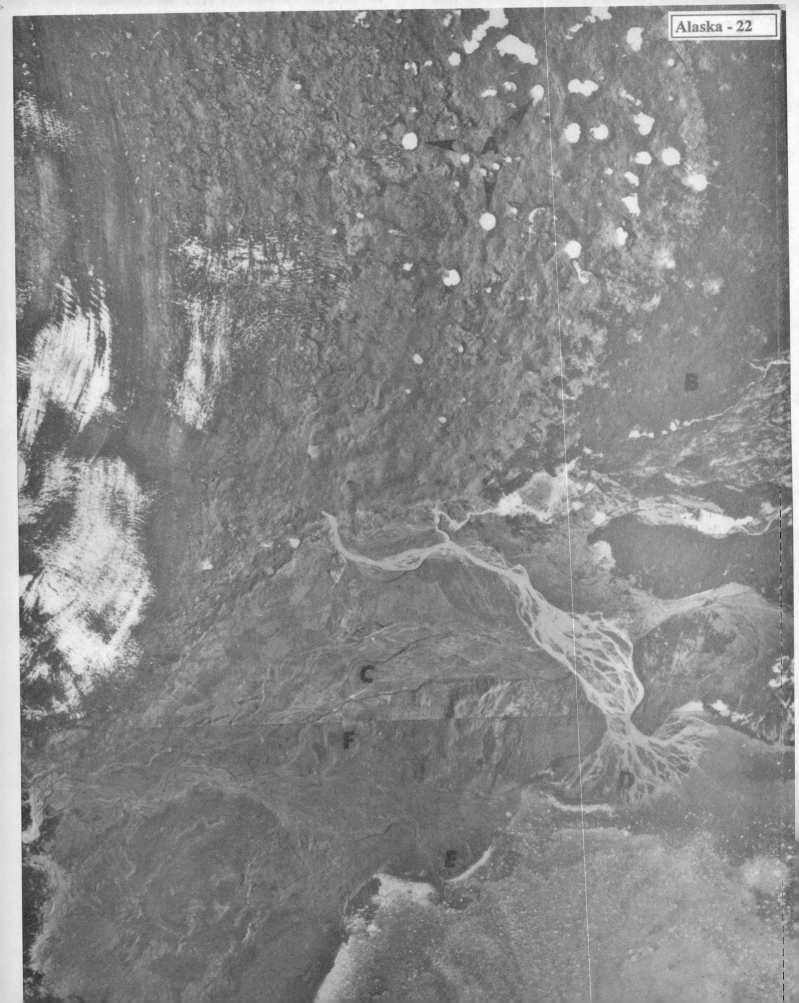

GLACIAL LANDFORMS – MT. SHUKSAN, WASHINGTON

1. Name the landforms at the following places:

a.   'A' _____

b.   Jagged Ridge at 'B' _____

c.   Summit of Mt. Shuksan (C) _____

d.   The valley of the Nooksack River (D) _____

e.   Price Lake (E) _____

f.   Shuksan Lake (F) _____

g.   Sulphide Creek where it flows out of Sulphide Lake (G) _____

2. The headwall of the East Nooksack Glacier rise about 2000' above the glacier. Did the glacier necessarily extend to the top of the headwalls of this cirque at one time? _____ How else might the headwall have been eroded so far above the present glacier?
_____

3. Why does the cirque at Sulphide Lake (G) have a glacier in it, whereas the cirque at Shuksan Lake (F) does not? _____

4.   If the elevation of the termini of cirque glaciers approximates the firn line, what is the elevation of the present firn line on this map? _____

5. Although some of the glacial features on this map are still in the process of forming, others are not and must have been formed during a period of more intense glaciation in the past. For example, what evidence on the map do you see that glaciers once extended much farther down valley than they do at present? _____.
How far did the longest valley glacier extend down valley from its source?
_____

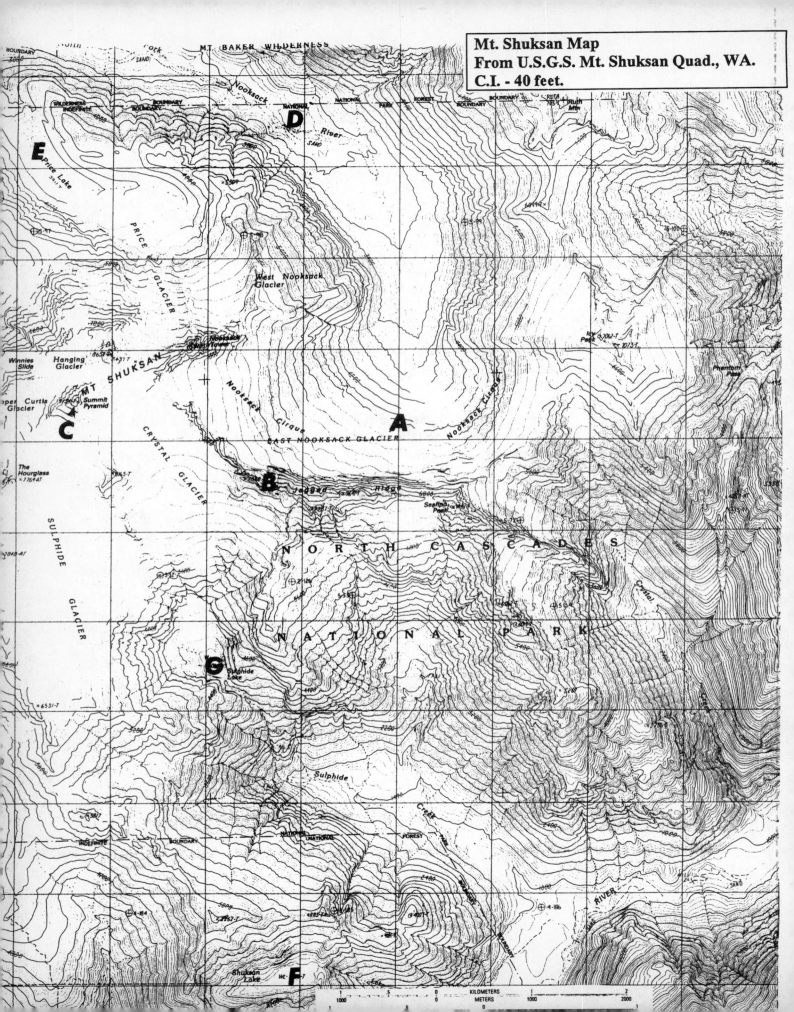

Mt. Shuksan seen from the south.

GLACIAL LANDFORMS – ROCKY MOUNTAINS, MONTANA

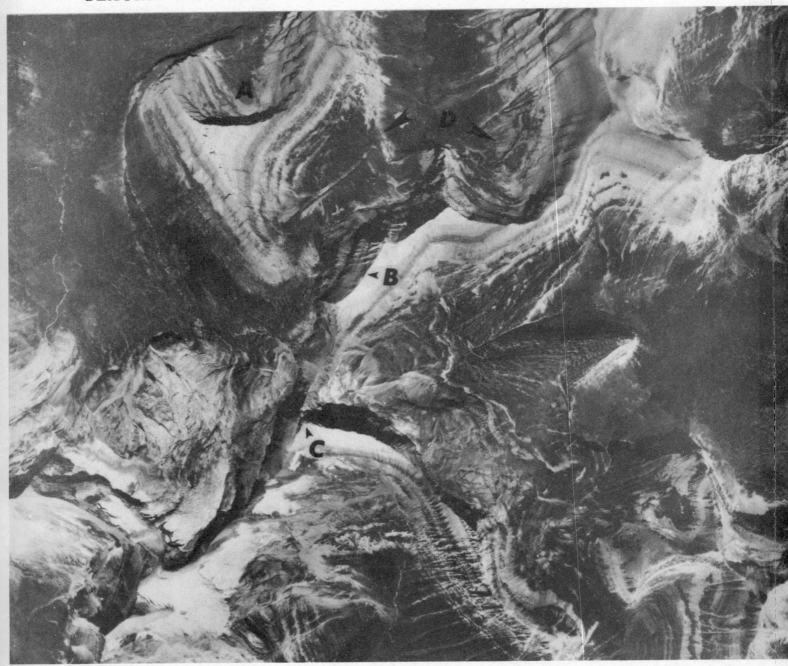

1. Name the landform at the following places: 'A' _____
'B' _____   'C' _____

2. Why is the cirque at 'D' so much larger than the cirque at 'A'? _____
How can you account for the large crescentric irregularities in its upper part?
_____

What name may be applied to this type of cirque? _____

GLACIAL LANDFORMS – HOLY CROSS, COLORADO

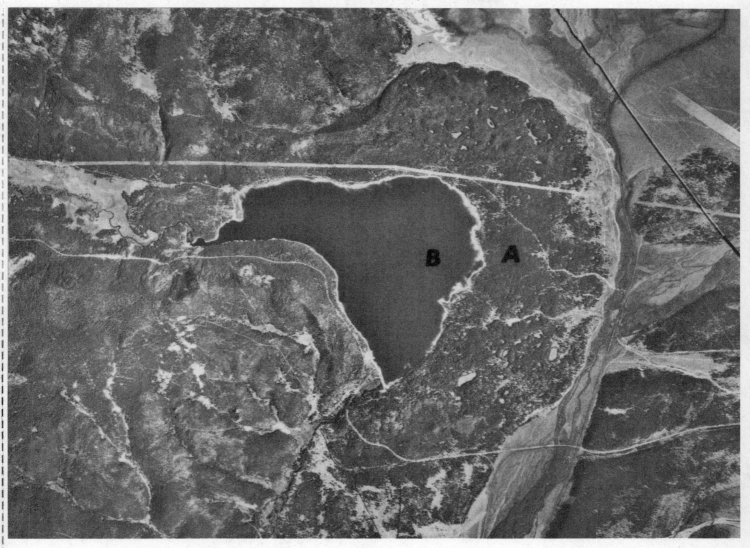

1. Name the landform at 'A' on the air photo and map. _____
How high is it above Turquoise Lake? _____ How wide is it? _____
What does the size of this feature suggest about the glacier that made it? _____

2. Name the landforms at the following places:

a. Turquoise Lake (B) _____

b. Lake Fork valley (C) _____

c. Mill Creek (D) _____

d. Valley of Glacier Cr. (E) _____

e. Galena Mt. (F) _____

f. Lonesome Lake (G) _____

Holy Cross Map
From U.S.G.S. Holy Cross Quad., CO.
C.I. - 50 feet.

GLACIAL LANDFORMS – MONO CRATERS, CALIFORNIA

1. Compare the linear ridges along the sides of Parker Creek Valley (A) on the topographic map with the same ridges shown on the air photo. Note the sharply crenulated contours making the ridges. Name the landform _____

2. Bloody Canyon (B), Lee Vining Creek (C), and Grant Lake (D) have similar ridges. Compare their topographic expression on the map with the air photo. Name the landform _____

3. How high above Parker Lake (A) are the linear ridges? _____ What does this tell you about the thickness of ice? _____.
How thick was the ice there? _____ How thick was the ice at Lee Vining Creek (C)? _____ How thick was the ice at Grant Lake (D)? _____

4. Why isn't a ridge of comparable height present around the lower end of Grant Lake? _____ How can you account for the difference in the size of end moraines here and on the Holy Cross map? _____

5. Name the landform that makes the peninsula into Grant Lake at 'E' _____
Can you tell whether this landform was caused by a recession of the ice or from a readvance of the glacier? _____

6. Name the landform making the ridges that border Sawmill Canyon (F). _____
Are they older or younger than the ridges bordering Bloody Canyon? _____ What is the evidence for you conclusion? _____

7. Name the landform at Kidney Lake (G) _____

8. Name the landform at Gibbs Canyon (from its headwaters to its mouth) _____

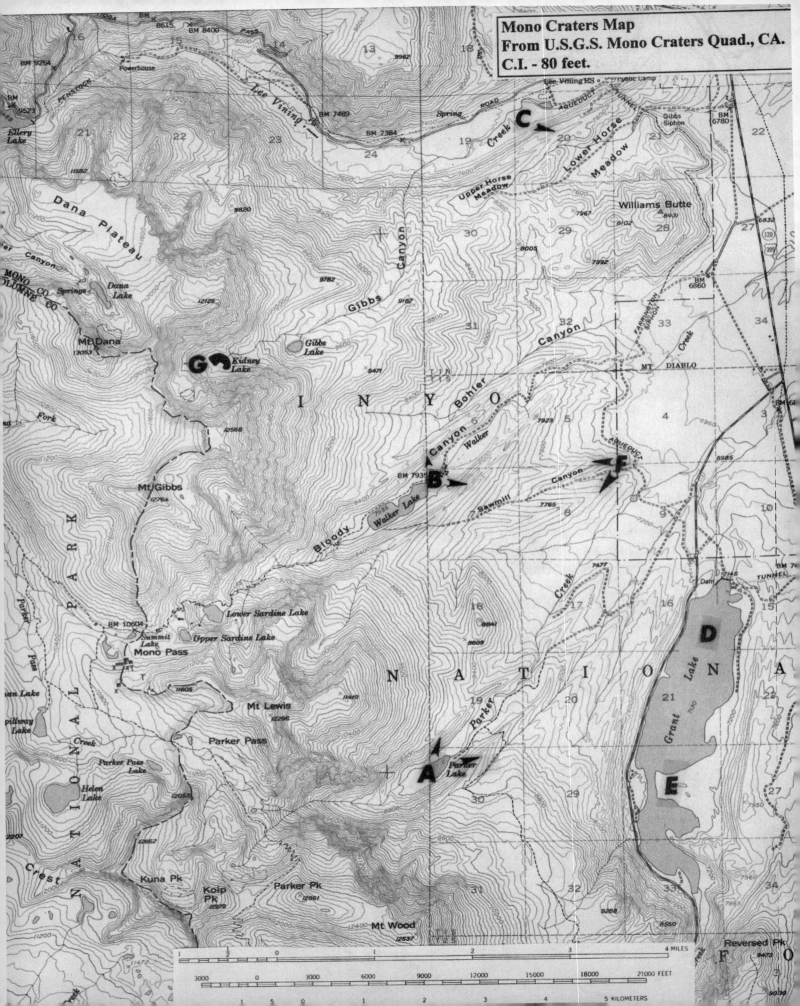

Mono Craters Map
From U.S.G.S. Mono Craters Quad., CA.
C.I. - 80 feet.

GLACIAL LANDFORMS – MONO CRATERS, CALIFORNIA

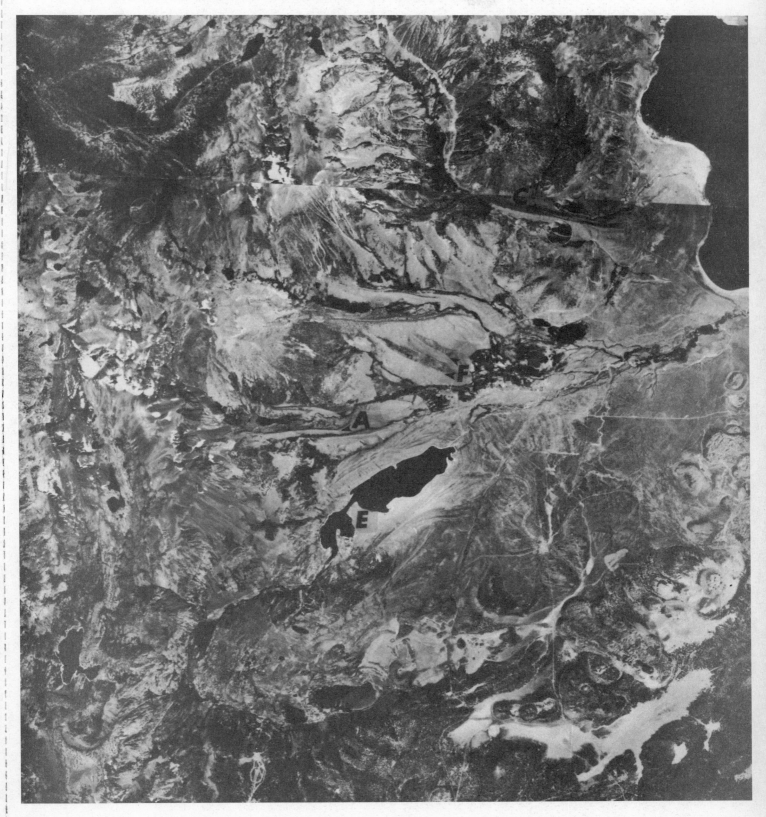

GLACIAL LANDFORMS – MONO CRATERS, CALIFORNIA

NAME _____

GLACIAL LANDFORMS – GRAND TETON, WYOMING

1. What landform holds in Jenny Lake (A)? (Compare the topographic map with the air photo).
_____

2. What landform holds in Leigh Lake (B)? _____

What landform holds in Taggart Lake (C)? _____

What landform holds in Bradley Lake (D)? _____

Why are the sizes of these landforms so different? _____

3. What landform holds in Jackson Lake (E)? _____
How does it differ in size and shape from Jenny and Leigh Lakes?
_____ Where was the source of the glacier that made
the landform that holds in Jackson Lake? _____

4. What landform makes up Burned Ridge and the topography at 'G'? _____
What is the relationship of this landform to the one holding in Jackson Lake (i.e., is it the same
age, same process)? _____
What is the evidence for your conclusion? _____

5. What is the origin of the Potholes (H)? _____

6. Name the landform that makes up Baseline Flat (I) _____

7. What is the origin of the shallow valley at 'J'? _____

Why is the channel dry? _____

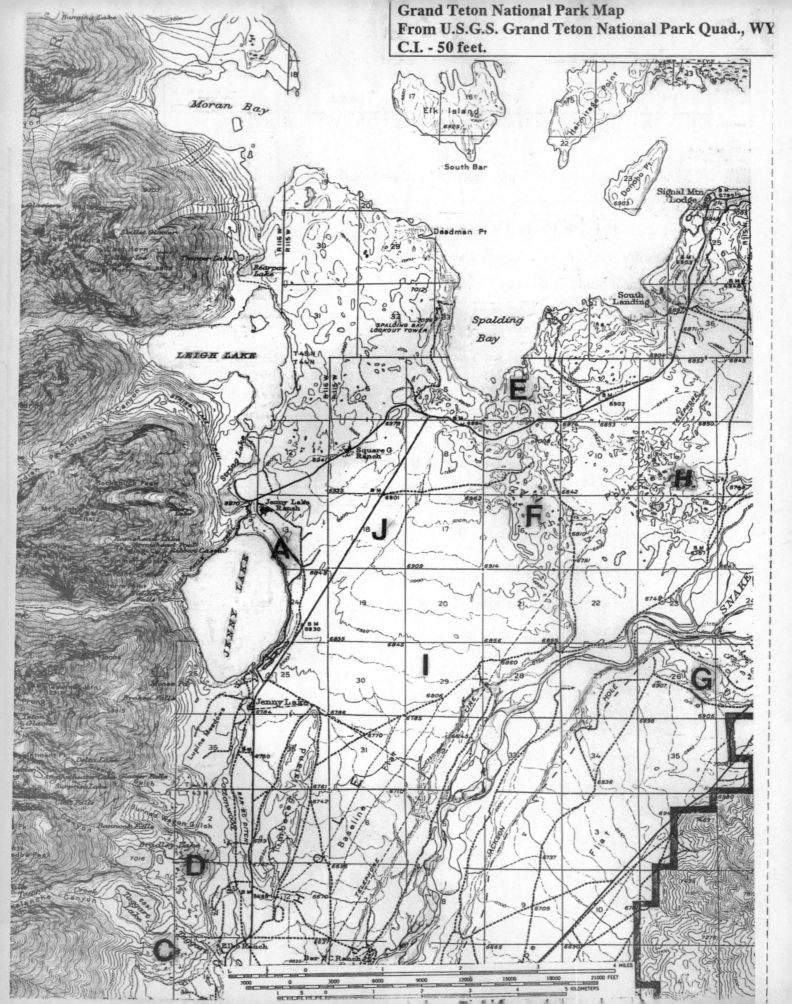

Grand Teton National Park Map
From U.S.G.S. Grand Teton National Park Quad., WY
C.I. - 50 feet.

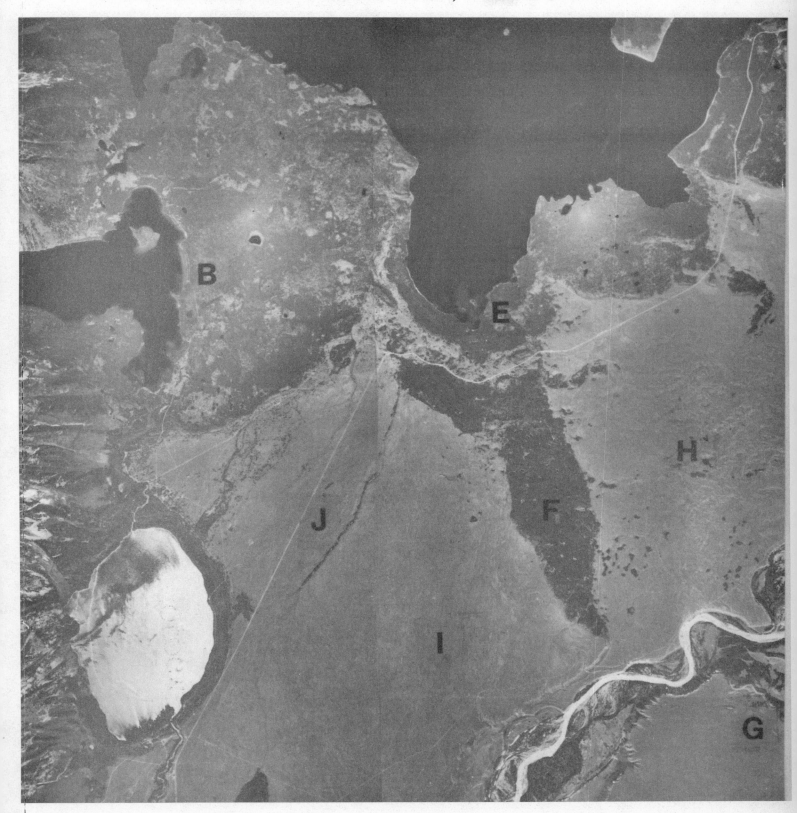

GLACIAL LANDFORMS – SOUTH DAKOTA

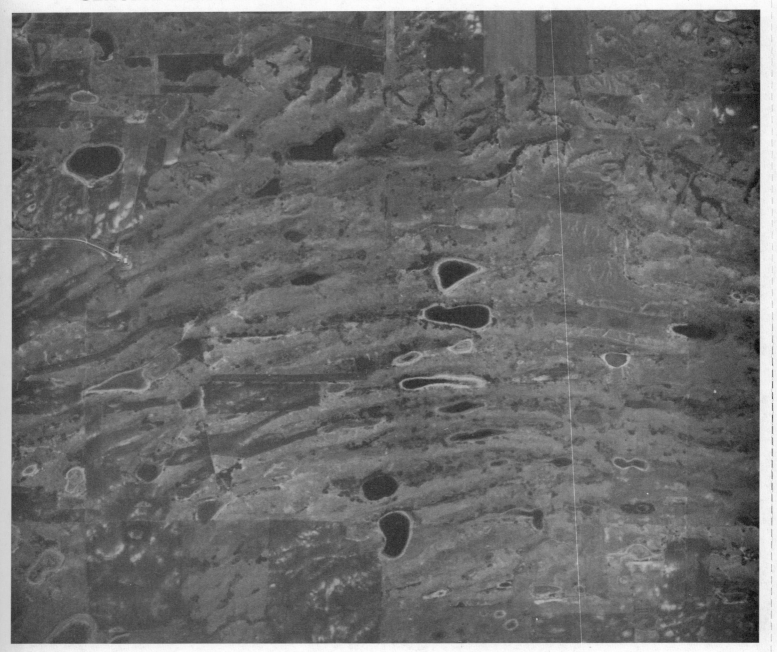

1. The ribbing on the moraine in this photo is caused by _____
which produces a _____ (landform).

2. Compare the difference in topographic expression with the moraine shown on the Kingston map and photo. How did the mass balances of the glaciers in these two areas differ?

_____

EXERCISE 13.7   NAME _____

GLACIAL LANDFORMS – KINGSTON, RHODE ISLAND

1.   The topography of the prominent ridge (A) is characterized by many hummocks and depressions, as shown by concentric contours. How wide is it? _____
Name   the   landform   _____   What   is   it   composed   of?
_____

2. The ridge contains several flat-topped hills at the left edge of the map (B). These landforms resulted   from   accumulation   of   sediment   in   _____   to   form
_____ (landform) Was the glacier active or stagnant when these hills were formed? _____

3. Name the landform making the ridges at 'C" on the air photo _____
How were they formed? _____

4. Name the landform making Factory Pond (D). _____
How was it formed? _____

5. Name the landform making the broad surface at 'E'. _____

6. Name the landform shown by the small circular, hachured contours at "F' (and the circular, dark areas on the air photo). _____

7. Name the landform at Green Hill (G) _____
What is the significance of its position relative to the ridge at 'A'? _____
Is it older or younger than the topography at 'A'? _____
What is the basis for your answer? _____

8. From the landforms on the map and photo, what can you infer about:
The direction of movement of continental glaciers in the area _____
The position of the margin of the glacier _____
The state of mass balance of the glacier that made Green Hill _____
The state of mass balance of the glacier that made the ridge _____

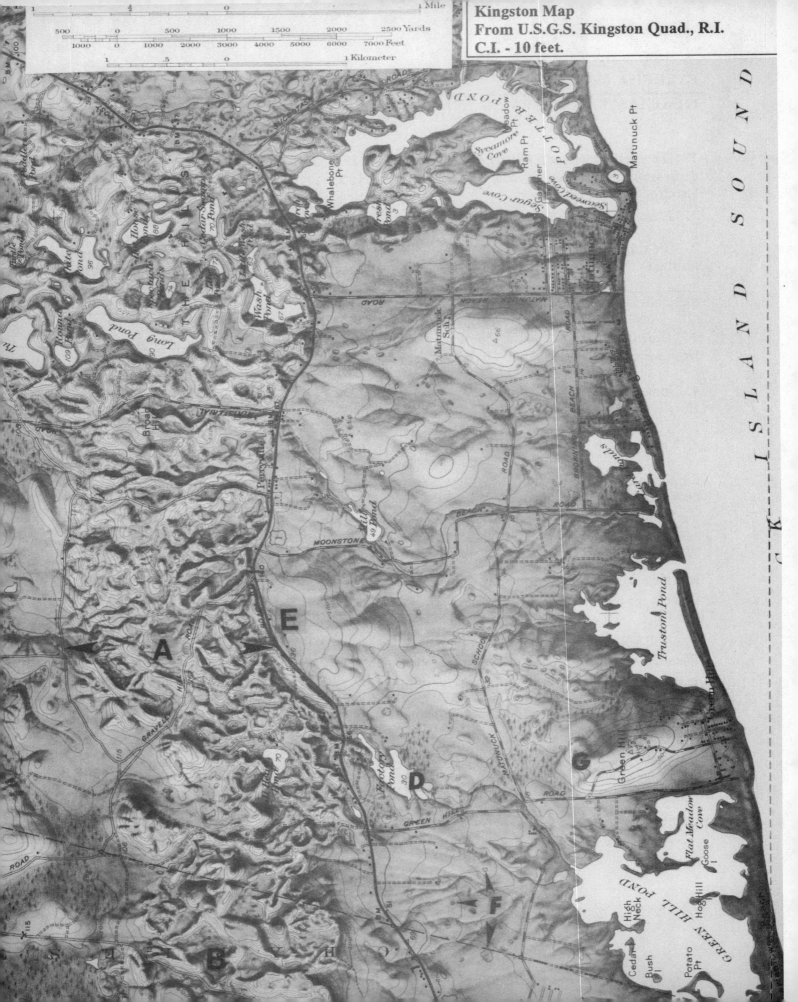

Kingston Map
From U.S.G.S. Kingston Quad., R.I.
C.I. - 10 feet.

GLACIAL LANDFORMS – KINGSTON, RHODE ISLAND

GLACIAL LANDFORMS – WHITEWATER AND WATERLOO, WI.

1. Name the landform making the elongate hills near 'A' _____

2. How is the form of the hills related to direction of ice movement? _____

3. What was the direction of ice movement here? _____

4. The Waterloo map shows a more detailed view of some of the elongate hills in this region. How long are they? _____ How wide are they? _____ The asymmetric shape of the hill at 'B' indicates the direction of glacier movement; the northern portion of the hill is steeper than the southern part and tapers in the _____

5. The form of these hills may have been shaped from pre-existing material, in which case they would be considered _____ in origin. How might you determine whether or not any particular hill is of this origin? _____

6. Can such elongate hills also attain their shape at the same time as the material in them is being deposited? _____ In this case, the origin would be _____ How might you determine whether or not any particular hill is of this origin? _____
_____

7. The topography making the ridge at 'C' is hummocky, with many depressions and closed contours. Name the landform at 'C' _____ What was its position relative to the ice when it was formed? _____

8. Name the landform that makes the flat surface at 'D' _____

9. How does the surficial material in the ridge near 'C' differ from that near 'D'? _____
Why? _____

10. If you wanted a source of well-sorted gravel to make concrete, where would you look?
_____

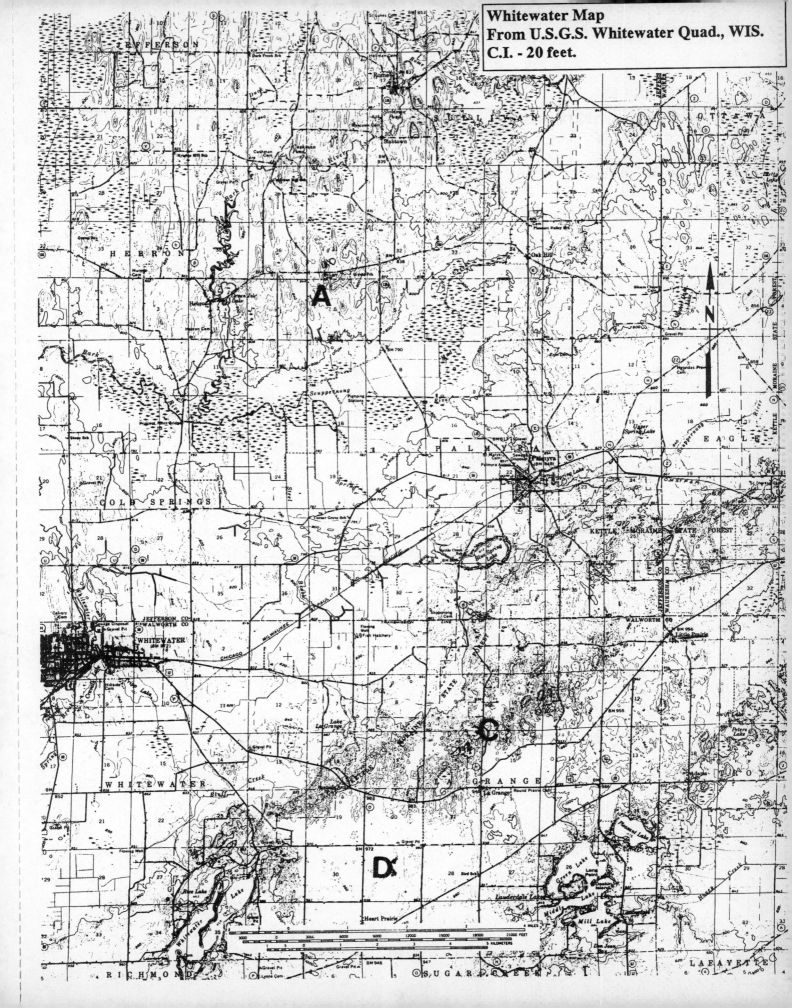

GLACIAL LANDFORMS – WHITEWATER AND WATERLOO, WI.

GLACIAL LANDFORMS – IVERSON, MINNESOTA

1. Name the landform that makes the prominent ridges at 'A' and 'B'. _____

What is the evidence for your answer? _____

What are the ridges composed of? _____Why do the ridges terminate

so abruptly? _____

2. Name the landform making the flat-topped hill at 'C' . _____

How did it form? _____

What does this tell you about the mass balance of the glacier? _____

3. Name the landform making the irregular hills at 'D' _____

4. What is the origin of Bob Lake (E)? _____

5. Name the landform making the NE-trending ridge between 'F' and 'G'. _____

What is its relationship to the ridges at 'A' and 'B'? _____

6. Name the landform making the NE-trending ridge between 'H' and 'I'. _____

What is its relationship to the ridge at 'F' - 'G'? _____

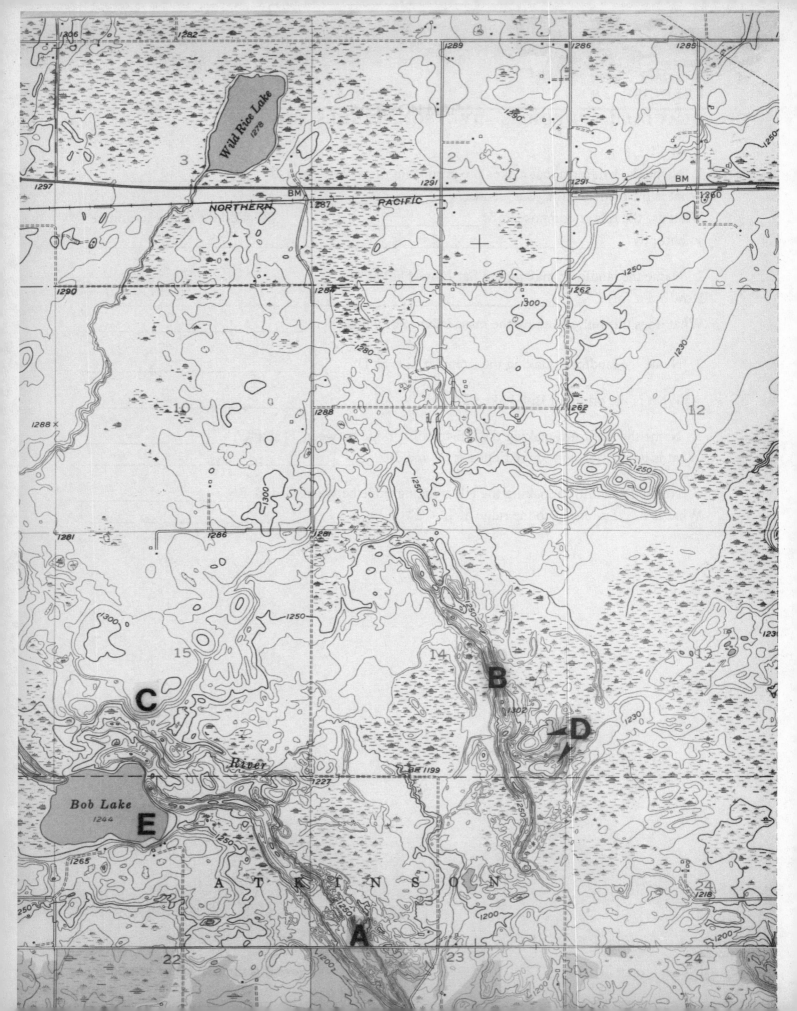

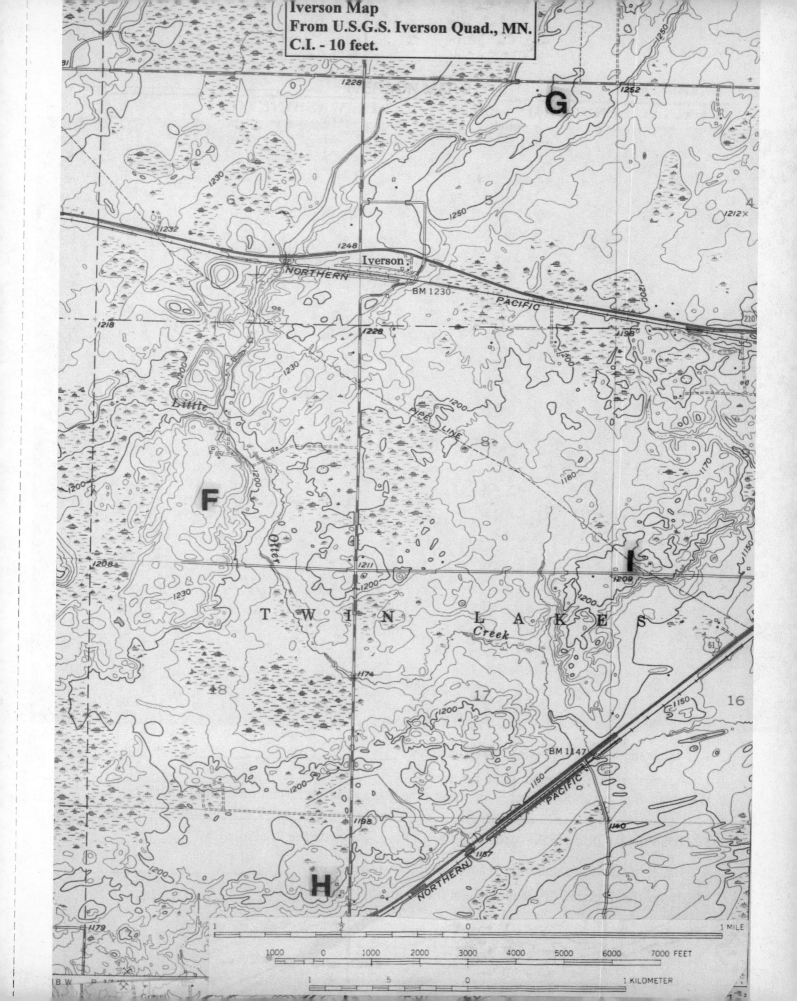

Iverson Map
From U.S.G.S. Iverson Quad., MN.
C.I. - 10 feet.

GLACIAL LANDFORMS – BARNES BUTTE, WASHINGTON

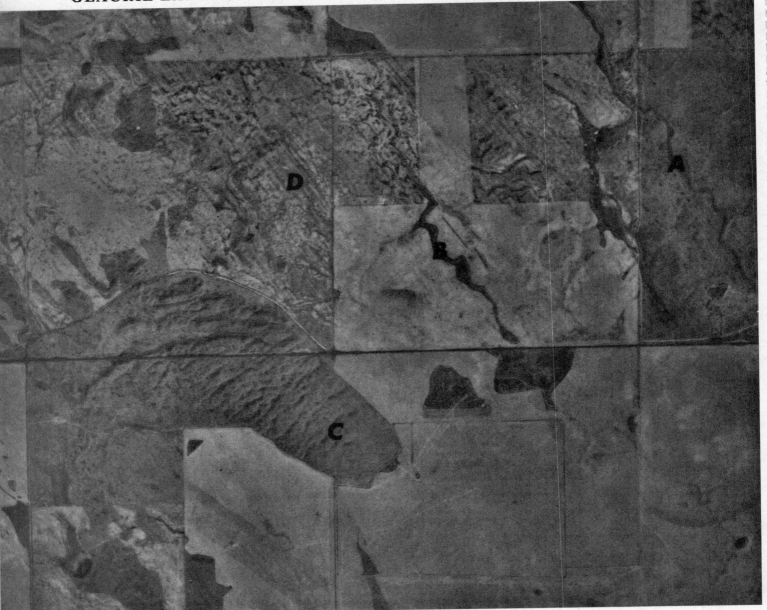

1. Names the landform making the ridge at 'A'. _____

2. Names the landform making the ridge at 'B'. _____
What can you day about the mass balance of the glacier during formation of these ridges?
_____

3. Name the landform making the large mound topography at the Pot Hills(C) _____

4. Name the landform making the subdued linear ridges at 'D'. _____

What was the direction of ice movement in this area? _____

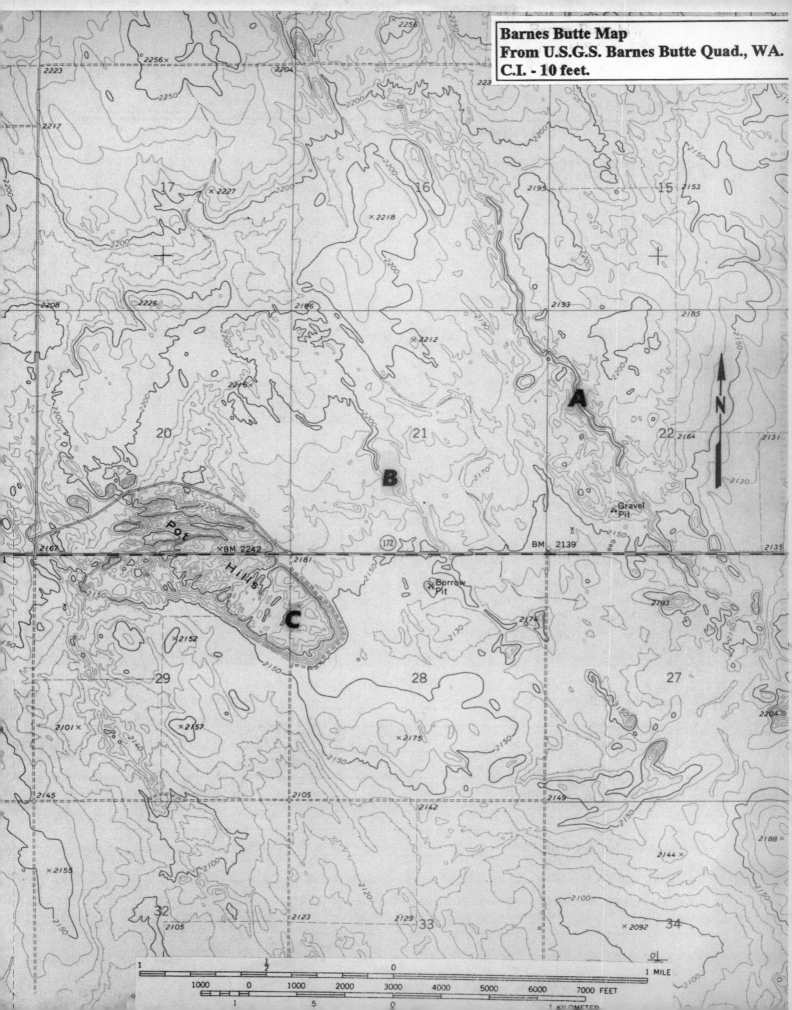

Barnes Butte Map
From U.S.G.S. Barnes Butte Quad., WA.
C.I. - 10 feet.

NAME _____

GLACIAL LANDFORMS – SOUTH DAKOTA

1. Name the landform at 'A' _____ How did it form?

_____

2. Name the landform at 'B' _____ How did it form?

_____

3. Name the landform at 'C' _____ How did it form?

_____

4. Name the landform at 'D' _____ How did it form?

_____

5. What can you say about the mass balance of the glacier that made these landforms?

_____

GLACIAL LANDFORMS – SIMS CORNER, WASHINGTON

1. Name the landforms at 'A' and 'B' _____

2. Name the landform at the distal end (C) of landform 'A' (compare the map with the photo).
_____ What are the linear features on the surface of 'C'
(shown on the photo)? _____ How is the landform at 'C' related to the
landform at 'A'? _____

3. Draw on the map the position of the ice margin when these landforms were forming.

4. What is the elevation of the surface of the landform at 'C'? _____
What is the elevation of the crest of the ridge leading to 'C' at the following places?

       'F' _____ 'A' _____ 'G' _____

How can you account for the fact that the ridge crest is lower than the landform at 'C' and has
several places with reverse slopes? _____

5. Name the landform at 'D' _____ How is it related to the landform at 'E'?
_____

6. What is the elevation of the landform at 'D'? _____ What is the elevation of the
ridge leading to 'D'? _____ How can you account for the fact that the ridge crest of 'E'
is lower than the surface of the landform at "D'? _____

7. What is the elevation of the crest of the ridge at the following places?

      'H' _____ 'I' _____ 'B' _____ 'J' _____

How can you account for the fact that the ridge crest reverse slopes at several places?
_____

8. Name the landform at 'K' _____

9. What do these landforms indicate about the vigor of ice movement when they were formed
(Was the glacier actively moving or was it stagnating?) _____

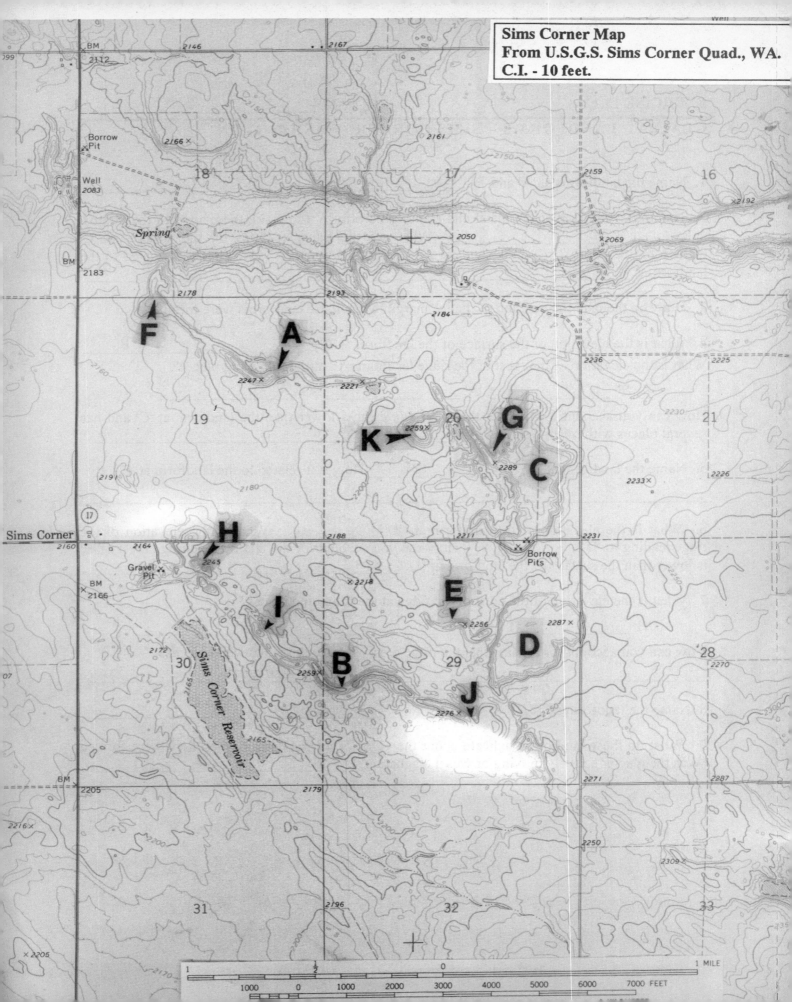

Sims Corner Map
From U.S.G.S. Sims Corner Quad., WA.
C.I. - 10 feet.

GLACIAL LANDFORMS – SIMS CORNER, WASHINGTON

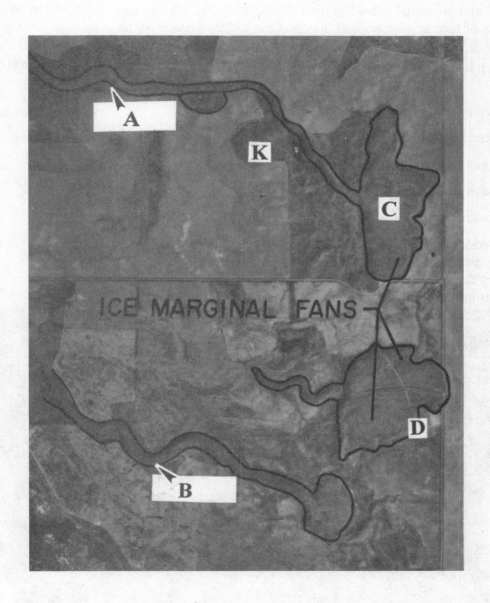

Pleistocene Climatic Changes and the Ice Ages

During the Ice Age, Extensive snow fields and alpine glaciers were prevalent in mountain ranges around the world. Many of the glaciers have now receded or melted away, but they leave depositional and erosional record of their former activity. Study of glaciated landscapes permits us to outline areas of former glaciers and interpret the glacial history of the area.

Topographic Criteria for Recognition of Pleistocene Climatic Changes

1. Variable elevations of cirque floors

2. Multiple end moraines

3. Eustatic and isostatic marine terraces

4. Pluvial lake shorelines

5. Glacial outburst flood features

    Abandoned channels
    Channels cut across stream divides
    Giant flood bars
    Giant gravel ripples

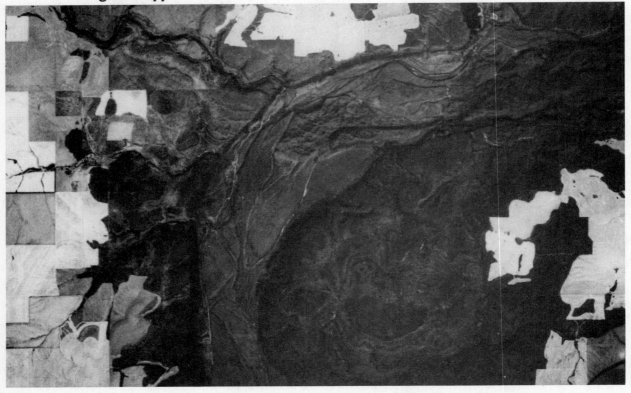

Moses Coulee, scoured by Missoula floods. The hummocky topography
in mid-valley is an end moraine draped over a giant flood bar.

150

NAME _____

PLEISTOCENE CLIMATIC CHANGES AND THE ICE AGES – WYOMING
FREMONT LAKE, WYOMING

Glaciers expanded and receded a number times during the Pleistocene. Many of the earlier glaciations were more extensive than later ones, leaving old moraines beyond the limits of the younger ones. The photograph shows a relatively fresh moraine (A) of the last major glaciation holding in the lake, surrounded by an older moraine (B).

1. Compare the photographic tone. Which is darker? _____

2. Compare the smoothness of the surface of the two moraines. Which is more subdued? Why?
_____

3. Note the small drainage channels on moraine 'B'. Do similar channels occur on moraine 'A'?
_____ Explain the reason for the difference. _____

_____

4. If you look carefully at the surface of moraine 'A' (using stereo-pairs or a magnifying glass) you can see large boulders on the surface (small light-toned dots). Do they also occur on moraine 'B'? _____ Because both moraines were made by glaciers in the same valley, they presumably had similar bouldery surfaces when they were formed. What has happened to the boulders that once must have lain on surface of moraine 'B'? _____

_____

PLEISTOCENE CLIMATIC CHANGES AND THE ICE AGES – WYOMING
FREMONT LAKE, WYOMING

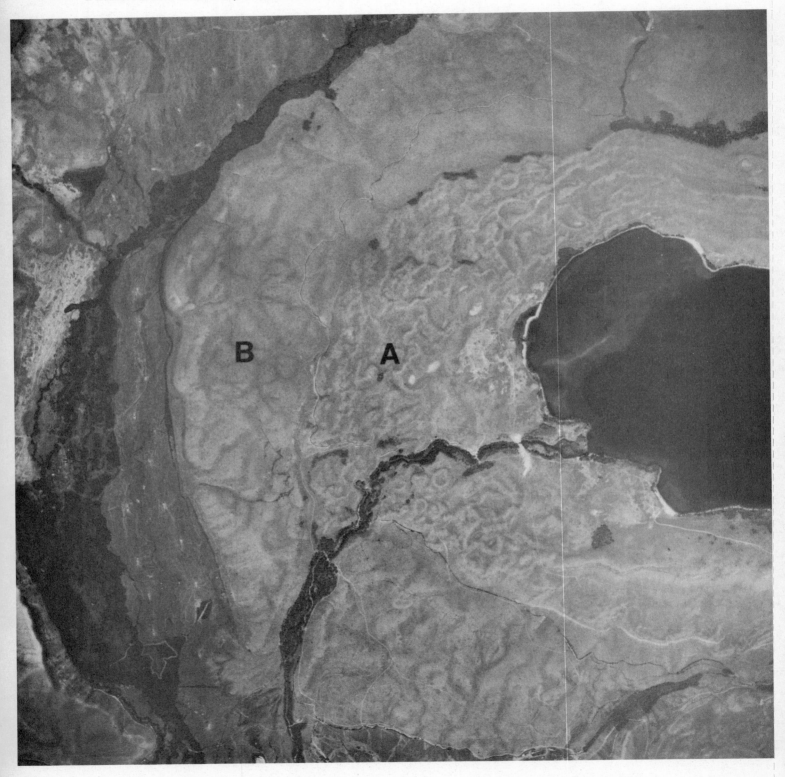

PLEISTOCENE CLIMATIC CHANGES AND THE ICE AGES –
MT. WHITNEY, CALIFORNIA

1. Kings-Kern Divide is a long, narrow ridge separating Kings Canyon National Park from Sequoia National Park. It is all that remains of a former high area that has been largely excavated by glacial erosion. The horseshoe-shaped basins eroded into the divide area are all _____ (landform). Back-to-back recession of the steep cliffs of these landforms produce many sharp-crested ridges known as _____ (landform) that now form Kings-Kern Divide. Spire-shaped peaks, such as Junction Peak, Mt. Stanford, Thunder Mt., Table Mt., among others, rise above the main level of the ridge to form _____ (landform).

2. The confluence of many smaller basins at 'V' form a large _____ (landform)

3. What is the elevation of Mt. Whitney (W)? _____ Mt. Muir (X)? _____
Mt. Russell (Y)? _____ Do any of them have active glaciers today? _____
What does this tell you about firn lines in this area? _____
Typical elevations of cirque floors in this area are about _____, suggesting that firn lines in this area during the Pleistocene were about _____ lower than today (subtract the elevation of the cirque floors from the presently unglaciated areas of Mt. Whitney (W), Mt. Muir (X), and Mt. Russell (Y).

4. Draw a topographic profile along the line Z - Z'.

Z   Z

Each one of the lakes lies in the floor of a _____ (landform) formed at successively higher elevations, suggesting that the firn line here has risen progressively from _____to _____ ft. and above.

5. Why does the gradient of Wallace Cr. steepen abuptly near Kern River? _____ Name the landform made by this steepening of the gradient. _____

6. Has the Kern River Valley been glaciated? _____ How can you tell? _____ How long was the glacier in Kern Valley? _____Why was the Kern River glacier so long? _____

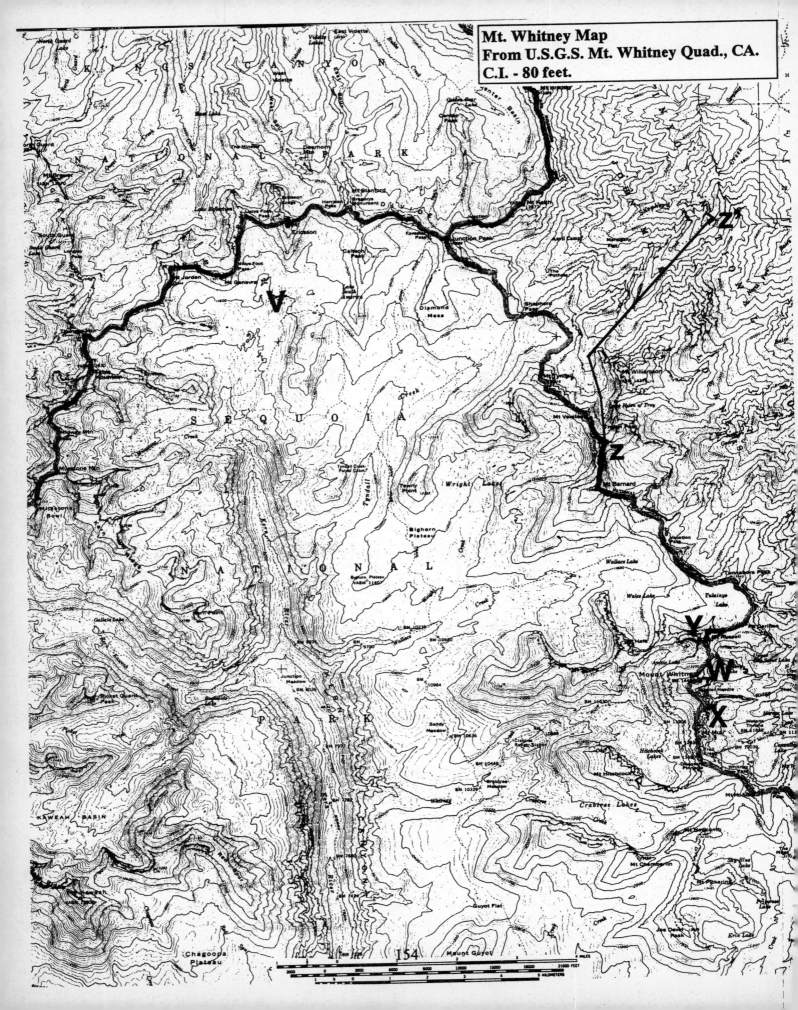

154

PLEISTOCENE CLIMATIC CHANGES AND THE ICE AGES –
PARK LAKE AND COULEE CITY, WASHINGTON

During the advance of continental ice sheets into northern Idaho and Montana, the Clark Fork River was dammed and a huge lake, Lake Missoula, was impounded. As the glaciers waned, the ice dam broke and sent gigantic floods across the Columbia Plateau in eastern Washington. These huge floods scoured immense volumes of basalt from the plateau and left large dry channels, known as *coulee,* the largest of which was Grand Coulee (A), and numerous waterfalls (B) and cataracts.

1. How deep is Grand Coulee (A) incised below the general level of the plateau? (Subtract the elevation of the lakes from the top of the cliffs at C. _____

2. Dry Falls (B) is the remnant of a large waterfall that extended from 'B' to 'D'. How wide was the falls? _____ How deep? _____

Note the anastomosing channels (E) that cover much of the map area. These channels were filled simultaneously and, in many places, cross divides between larger channels. Note the scoured topography at 'F'. This type of topography led to the term *channeled scablands.*

3. Giant gravel bars are common in many of the channels, as the one at 'G'. How long is the giant bar at 'G'? _____ How high is it? _____ Could such a bar be deposited by a normal river? _____

4. Additional evidence for the magnitude of the Missoula floods is found in giant ripples (H, I) composed of cobbles and boulders, such as the ones in the photos. Could such ripples be deposited by a normal river? _____ Why or why not? _____

    _____

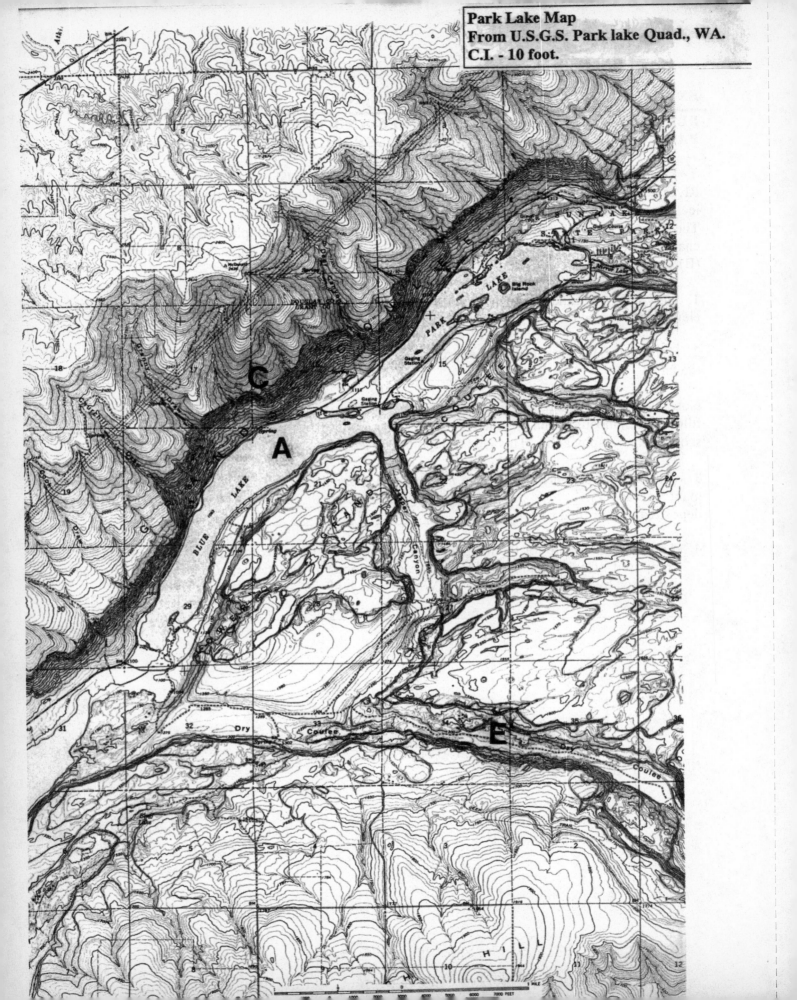

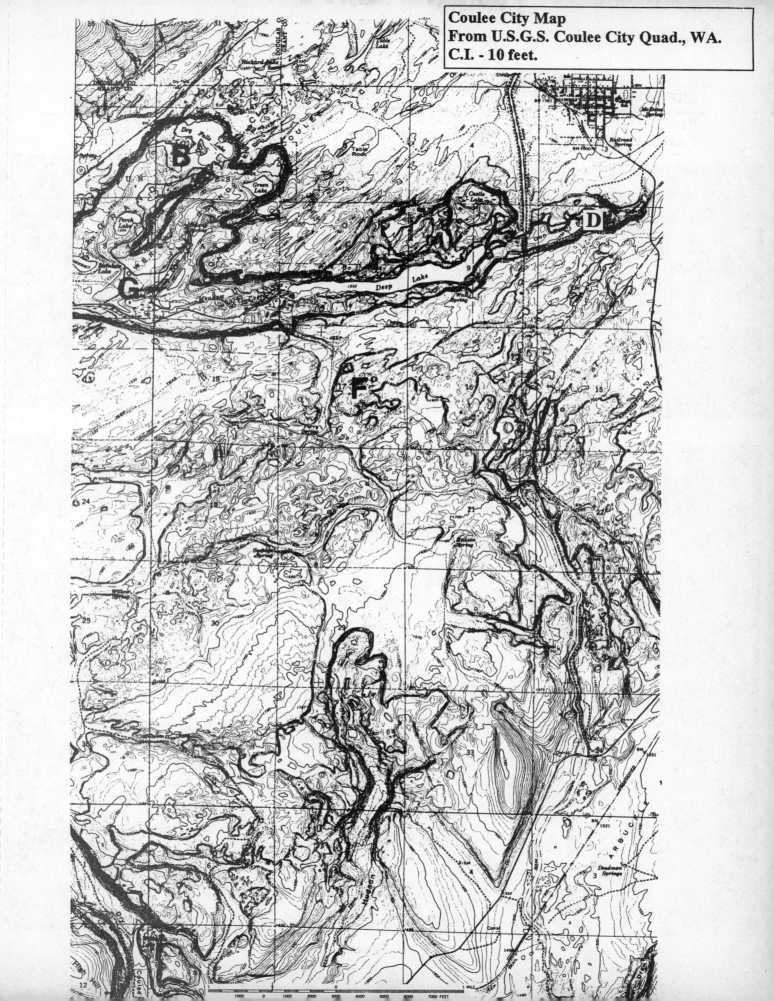

Coulee City Map
From U.S.G.S. Coulee City Quad., WA.
C.I. - 10 feet.

PLEISTOCENE CLIMATIC CHANGES AND THE ICE AGES –
PARK LAKE AND COULEE CITY, WASHINGTON

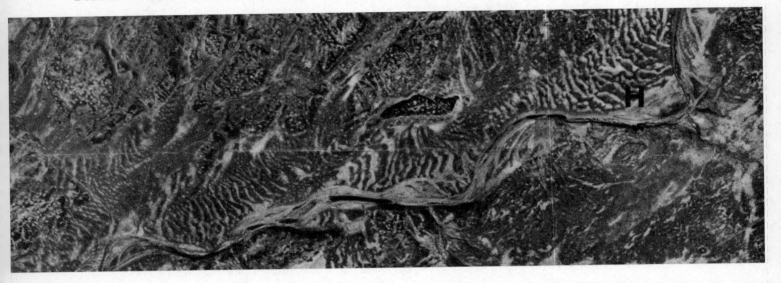

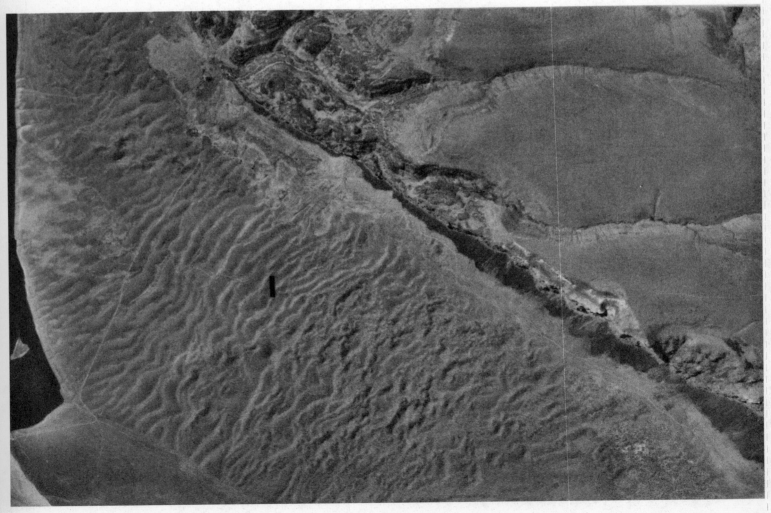

PLEISTOCENE CLIMATIC CHANGES AND THE ICE AGES –
JAMESON LAKE, WASHINGTON

Moses Coulee is a Missoula-flood-scoured, deeply incised, dry valley, much like Grand Coulee.

1. How deeply incised in the coulee? _____ How wide is it? _____
Does any stream now flow in it that is capable of eroding such a deep valley? _____

2. Name the landform at A' _____ What is it composed of? _____
Is it older or younger than Moses Coulee? _____ How can you tell?
_____

What does this imply about the age of the glacier responsible for releasing the flood waters that carved Moses Coulee? _____

3. Name the landform at 'B' _____ What are the faint sinuous features
on its surface? _____ What kind of material would you expect to
compose the landform? _____

4. The ridge at 'C' is composed of the same bedrock as the rest of the valley sides. The ridge at 'D' is composed of cobble gravel. How high above the coulee floor is the crest of the ridge at D? _____ Its composition (cobble gravel) and its position on the inside of a bend, in the coulee, in the lee of bedrock of the former valley side, suggest that it is a _____ of the Missoula flood.

5. Note the relationship of the landform at 'A' relative to the landform at 'D' (best seen on the photo). Which is older, the landform at 'A' or the landform at D? _____
How can you tell? _____

PLEISTOCENE CLIMATIC CHANGES AND THE ICE AGES – JAMESON LAKE, WASHINGTON

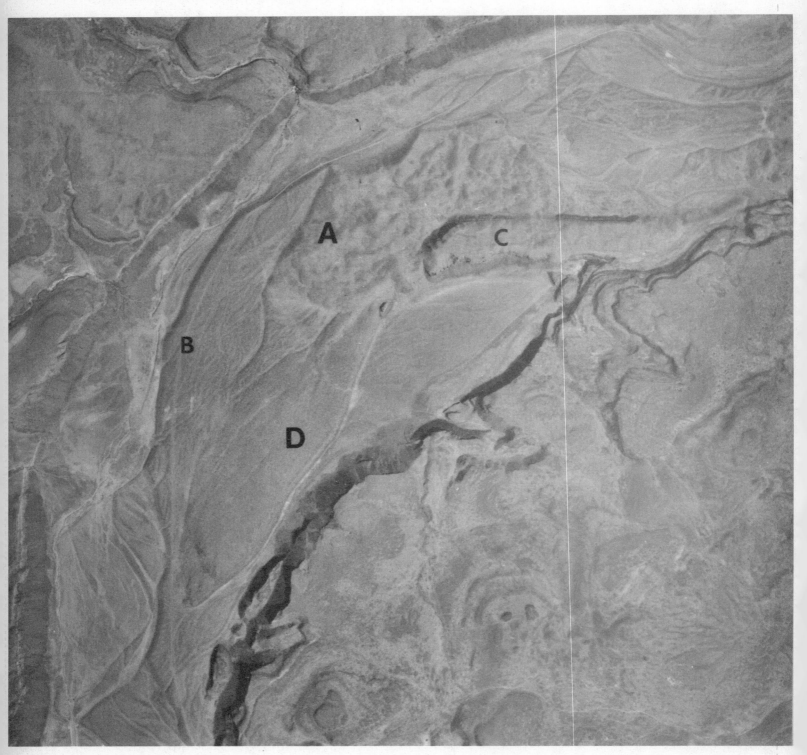

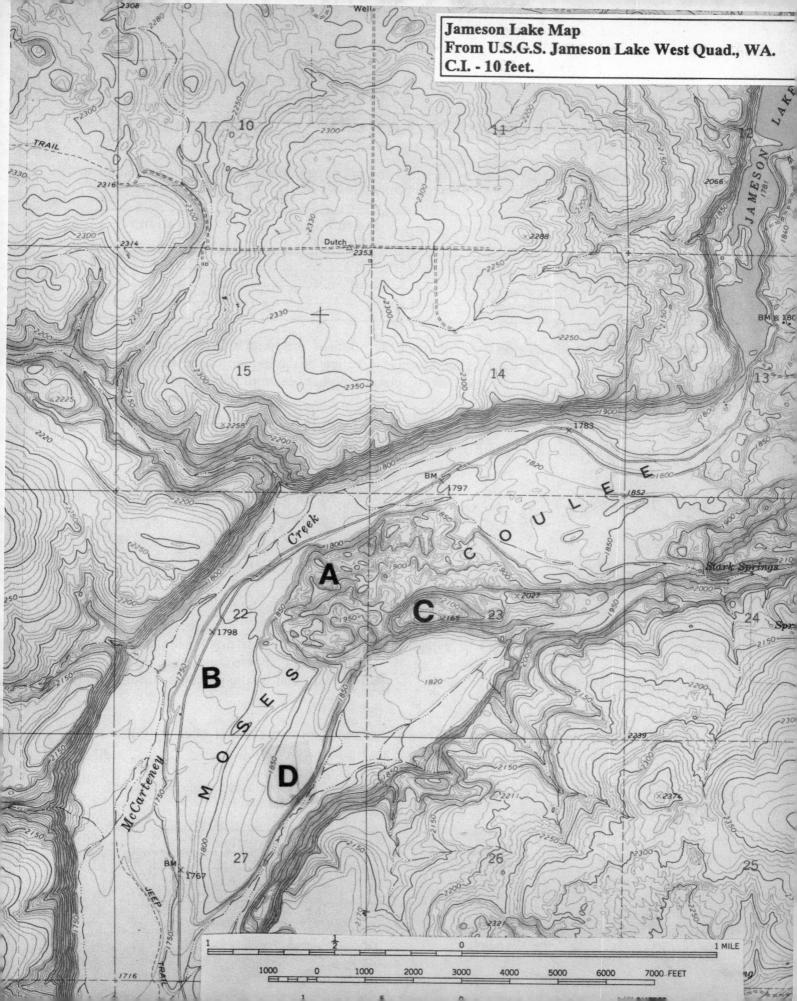

Jameson Lake Map
From U.S.G.S. Jameson Lake West Quad., WA.
C.I. - 10 feet.

NAME _____

PLEISTOCENE CLIMATIC CHANGES AND THE ICE AGES – HOLDEN, WASHINGTON

1. The firn lines may be approximated by the elevation of cirque glaciers' termini (but not valley glaciers) What is the elevation of the termini (approximate firn line) of the glaciers listed below?

North Facing Glaciers	Elevation	South Facing Glaciers	Elevation
Lyman (A)		Isella (E)	
Fortress Mtn (B)		Mary Green (F)	
High Pass (C)			
Buck Mtn. (D)			

2. Although some variation exists, the average elevation of the termini of the north-facing glaciers is _____ (ft.)whereas the average elevation of the termini of the south-facing glaciers is _____ (ft.). Do you see any relationship between terminus elevation and the direction the glaciers face? _____ What is the relationship?

_____

3. The diagram below shows the angle of the sun's rays relative to N-facing and S-facing slopes.

Which slopes receive the sun's rays most directly (i.e., at the highest angle)? _____
Which slopes are more shaded in the afternoon? _____ Because of these factors, _____-facing slopes have higher melting rates and thus firn lines and glacier termini are _____ (higher or lower?). For example, compare the elevations of the termini of the north-facing Company Glacier _____with that of the nearby south-facing Isella Glacier _____.

Another example occurs near Mt. Maude where the terminus of the Entiat Glacier (G) is at _____ whereas the cirques at Ice Lakes (H) are ice-free at elevation _____

4. During the Pleistocene, firn lines were considerably lowered. Note the elevation of cirque floors at the following places:

Glacier	Elevation
Lyman Lake (I)	
Holden Lake (J)	
Cuyuse Lake	

How much lower were firn lines here during the Pleistocene? _____

NAME _____

PLEISTOCENE CLIMATIC CHANGES AND THE ICE AGES – HOLDEN, WASHINGTON

5. Name the landform at the following places:

Bonanza Peak (K) _____

The peak at (L) _____

Valley of Railroad Creek (N) _____

Holden Lake (J) _____

6. Draw a topographic profile along the line P-P'.

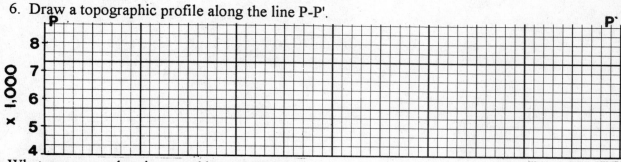

What name may be given to this topography? _____
What are the three possible origins of the alternating steep slopes and benches shown in the profile?

a. _____

b. _____

c. _____

7. If the Lyman Glacier (A) were to melt completely away, a bench would appear, very similar to the ones shown on your profile. If each of these benches is a cirque floor, what can you say about former firn lines here? _____

8. Draw topographic cross-valley profiles of the Napeeque River along the lines Q-Q' and R-R'. What is the reason for the difference in profiles? _____

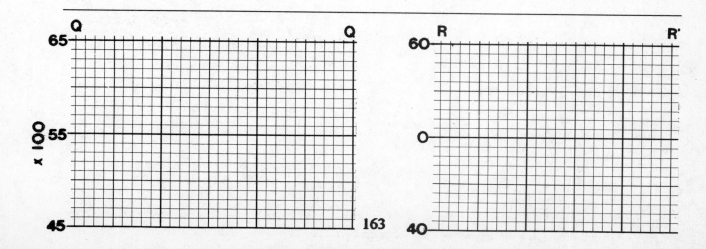

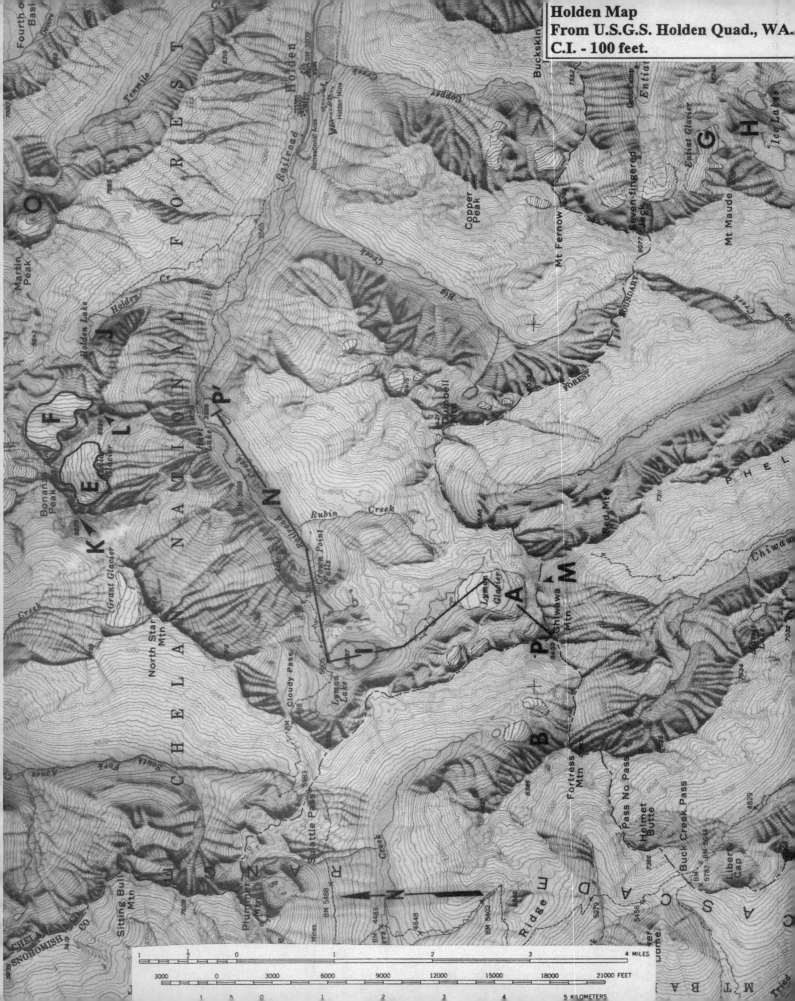

Holden Map
From U.S.G.S. Holden Quad., WA.
C.I. - 100 feet.

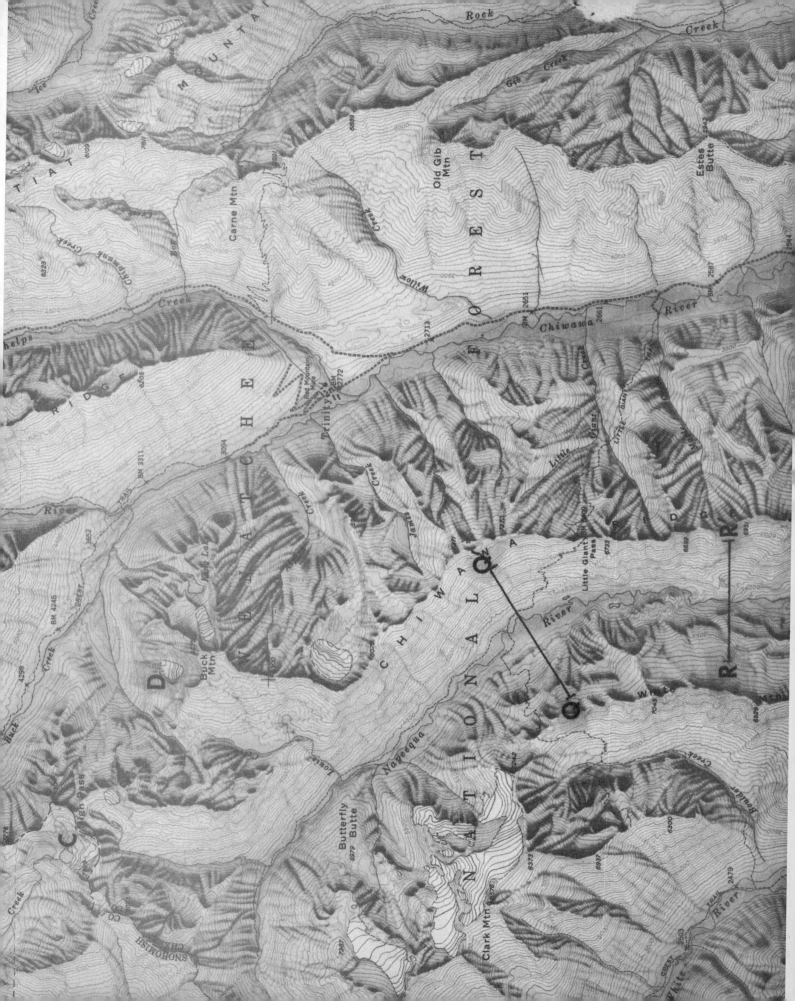

Shorelines

Topographic Criteria for Recognition of Shorelines

1. Coastal erosion
   - Sea cliffs
   - Wave-cut platforms

2. Coastal deposition
   - Beaches, beach ridges
   - Spits
   - Bars
   - Barrier island and bars

3. Submergent coastlines
   - Drowned valleys
   - Deep embayments
   - Numerous islands
   - Irregular coastlines

4. Emergent shorelines
   - Straight coastlines
   - Marine terraces

5. Coral reef development
   - Fringing reefs
   - Barrier reefs
   - Atolls

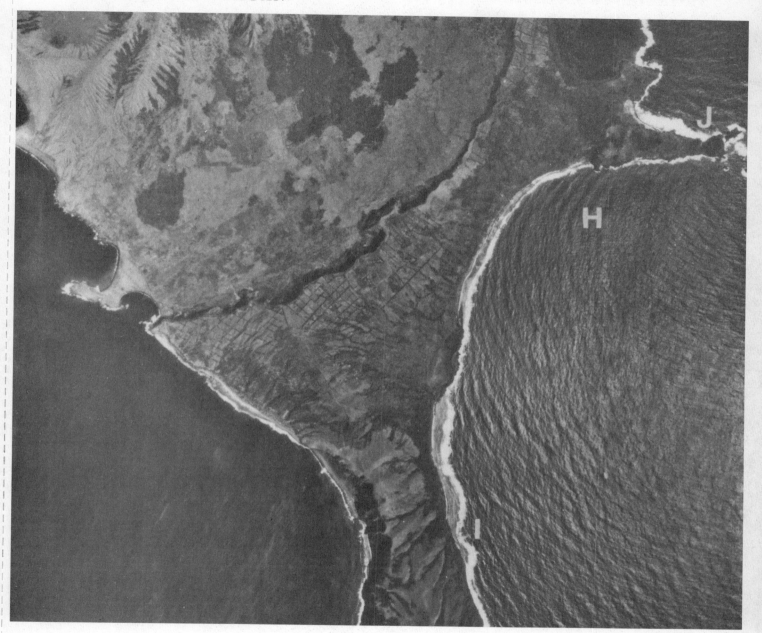

1. As waves approach the shore between 'H' and 'I', why do they bend as they approach the shoreline? _____

What is this process called? _____

2. Draw lines at right angles to the wave crests (orthogonals). Where does the wave energy converge? _____ Where does it diverge? _____

3. Why is the shore at 'J' bare and rocky, whereas the shore at 'I' consists of beach sediment? _____

4. Show the direction of longshore sediment transport between 'J' and 'I' with arrows.

167

SHORELINES – ALASKA

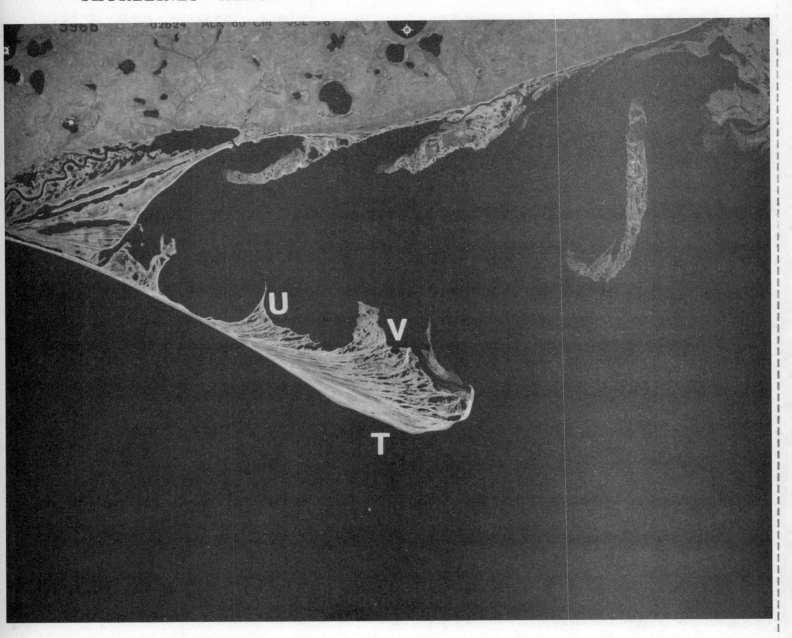

1. Name the landform at 'T'. _____

2. Show the direction of longshore sediment transport with arrows. _____

3. What is the origin of the linear ridges on the landform? _____

4. Which of the ridges is oldest, 'U' or 'V'? _____

5. Describe the evolution of the landform. _____

SHORELINES – ST. VINCENT ISLAND, FLORIDA

1. Name the landform shown. _____
2. What is the origin of the linear ridges? _____
3. Which of the ridges is the oldest, 'M' or 'N'? _____

Plunging breaker approaching the shoreline

NAME _____

SHORELINES – CHATHAM, MASSACHUSETTS

1. Name the landform making up Nauset Beach (A). _____

2. Note the underwater sand ridge at 'B' just offshore and another just to the south that precedes it. These submarine bars are pushed by waves around the end of the landform, making discrete hooked ridges at the distal end of the landform. Mark (on the map) places where this has happened in the past during evolution of the landform.

3. In the past 100 years the distal end of the landforms as grown southward about 2.5 miles. Mark on the map the approximate position of the end of the landform 100 years ago.

4. What has been the rate of southward growth in the past 100 years? _____

5. Has the shore at Chatham Lighthouse (C) ever been open to the ocean? _____
Using the southward rate of growth of the distal end of the landform, about when would it have become protected from the open ocean by Nauset Beach? _____

Chatham Map
From U.S.G.S. Chatham Quad., MASS.
C.I. - 10 feet.

SHORELINES – EDGARTOWN, MASSACHUSETTS

This map consists of a combination of shoreline and glacial features.

1. Suppose you happen to be out for an evening drive on Chappaquidick Island and decide to take a do-it-yourself field trip. As you drive along Dyke Road (D), you notice that you are driving along a _____ (landform) just north of the road, based on your observation of _____ What kind of material would you expect to see in the roadcuts? _____

2. If you fancied a stroll on the beach and drove eastward along Dyke Road from Tom's Neck across the bridge at (E) you would arrive at a _____ (landform).

3. If you decided to take a midnight swim from Chappaquidick Point (F) to Edgartown (G), rather than taking the ferry, you would be swimming from a _____ (landform) to a _____ (landform).

4. The topography west of Edgartown (H) has many depressions. What are the two possible origins of these depressions? _____ _____
Which of the two is the true origin? _____ What evidence is the basis for your answer? _____

5. Name the landform making up the topography at (I) that extends south of Edgartown to the sea _____ What is the evidence for you answer? _____

What kind of material would you expect to find there? _____

6. The many inlets and coves of Edgartown Great Pond (J) formed as a result of _____ of the coastline.

7. Name the landforms at the following places: a. Norton Point (K) _____

b. Note the dots making up eastern Norton Point. They represent new deposition between the 1972 and 1977 editions of the map. In 1972, Norton Point was a _____ (landform) that was made into a _____ (landform) between 1972 and 1977.

c. Between Wasque Point (L) and Cape Poge (M). _____ What is the direction of longshore transport? _____

d. Cape Poge Elbow (N) _____ What is the direction of longshore transport? _____ Edgartown Beach (O) _____

f. What is the direction of longshore transport between Oak Bluffs (P) and Edgartown Beach (O)? _____ What is the *evidence* for your answer? _____

_____

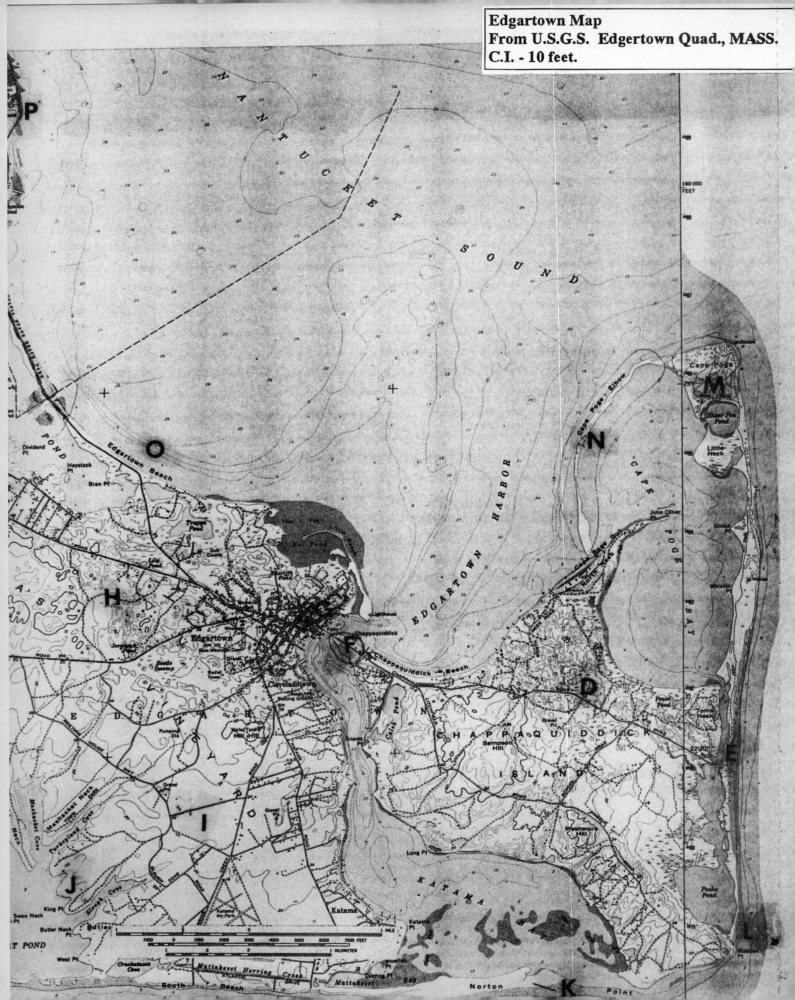

Edgartown Map
From U.S.G.S. Edgertown Quad., MASS.
C.I. - 10 feet.

EXERCISE 15.5
NAME _____

SHORELINES – TISBURY GREAT POND, MASSACHUSETTS

1. Tisbury Great Pond (A) has many estuaries that extend northward as a result of
_____ of the coastline here.

2. Name the landform making up the barrier (B) across Tisbury Great Pond. _____

3. Name the landform making up the barrier across Watcha Pond (C). and Oyster Pond (D).
_____

4. The reason for the present straightness of the shoreline on the map is
_____

EXERCISE 15.6

SHORELINES – PALOS VERDES HILLS, CALIFORNIA

1. Draw a topographic profile along the line A-A'.

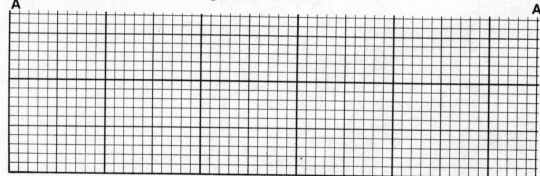

2. What is the origin of the benches in the profile? _____

3. How high is the highest bench? _____ What does this suggest about the cause of the
benches? _____

4. How many benches can you identify? _____ What does this suggest about the
history of the coast here? _____

175

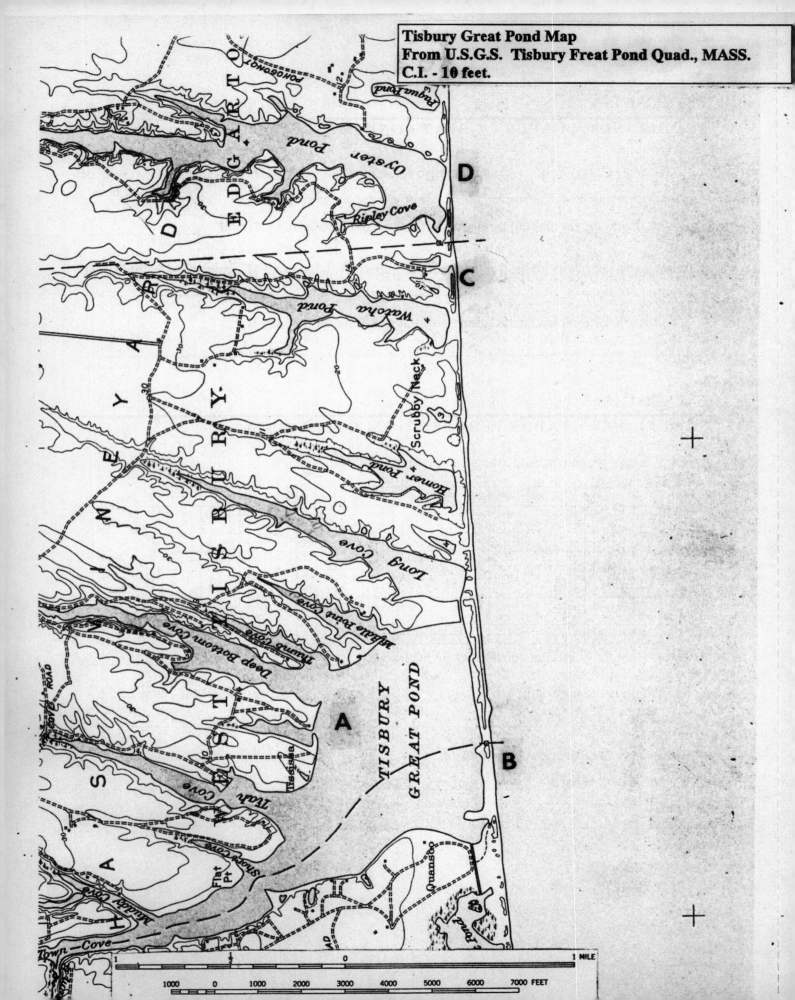

Tisbury Great Pond Map
From U.S.G.S. Tisbury Freat Pond Quad., MASS.
C.I. - 10 feet.

SHORELINES – PALOS VERDES HILLS, CALIFORNIA

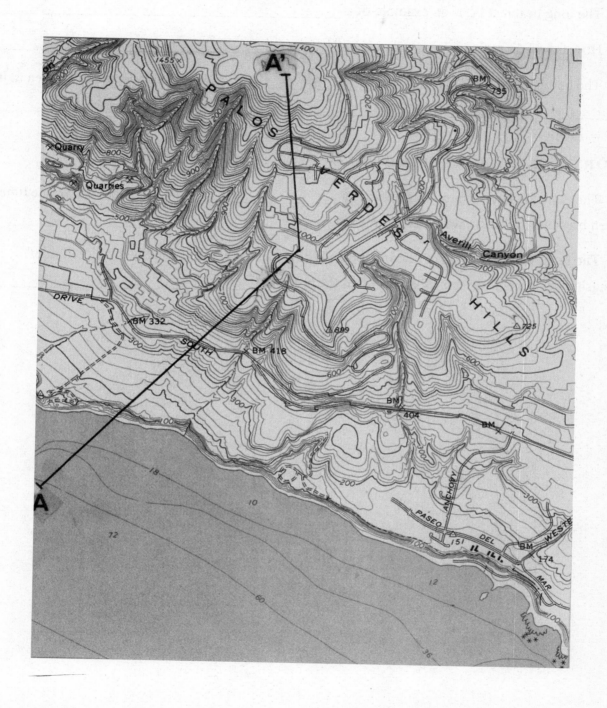

NAME _____

SHORELINES – MELBOURNE, FLORIDA AND NORTH CAROLINA

1. The long beach at 'X' is an example of a _____

2. How far offshore from the mainland is it? _____

3. The body of water at 'Y' is a _____ What will eventually happen to it?

_____

NORTH CAROLINA PHOTO

1. The elongate landform on this photo is a _____ that has at one time
been breached by storm waves at 'J', forming a _____

2. The feature at 'K' has since filled in most of the breached beach. What is the direction of
longshore sediment transport that has filled the breach? _____

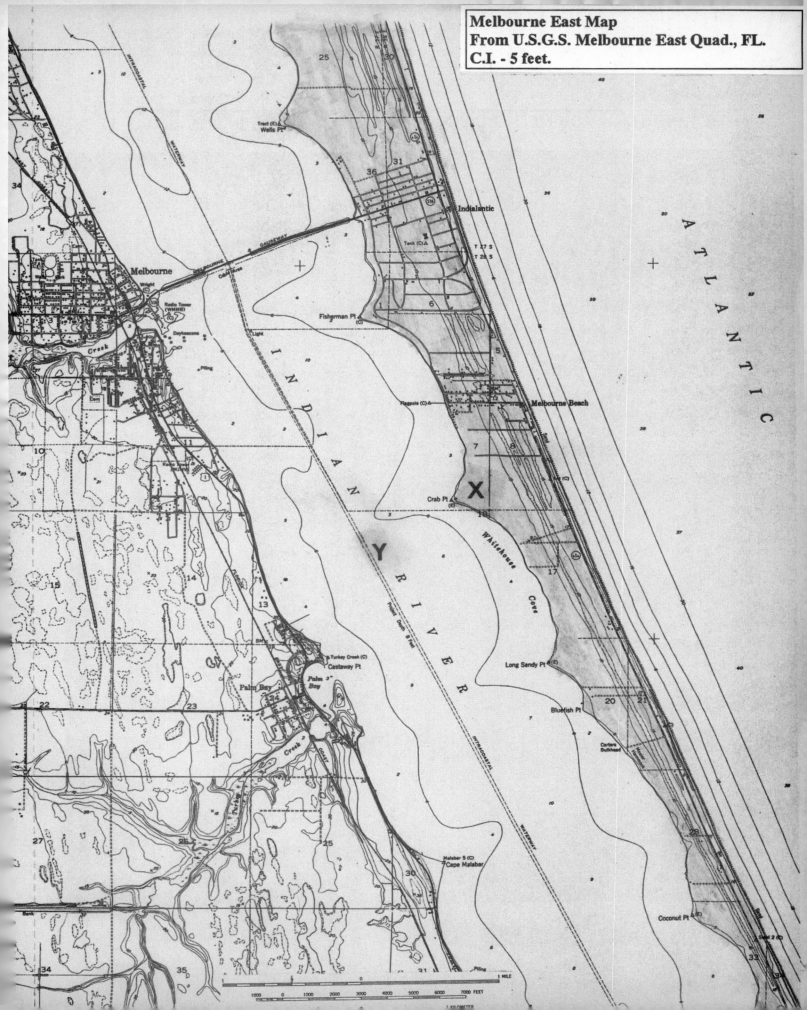

SHORELINES – NORTH CAROLINA

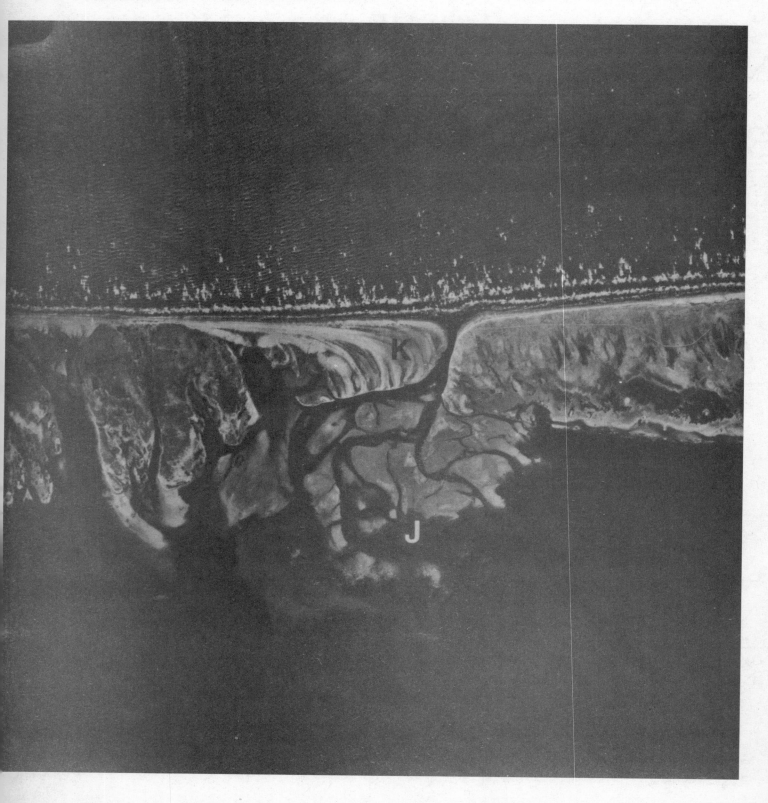

Eolian Processes and Landforms

Eolian Features

1. Dune Types

> Barchans
>
> Parabolic dunes
>
> Transverse dunes
>
> Longitudinal dunes
>
> Star dunes

EOLIAN PROCESSES AND LANDFORMS – CALIFORNIA

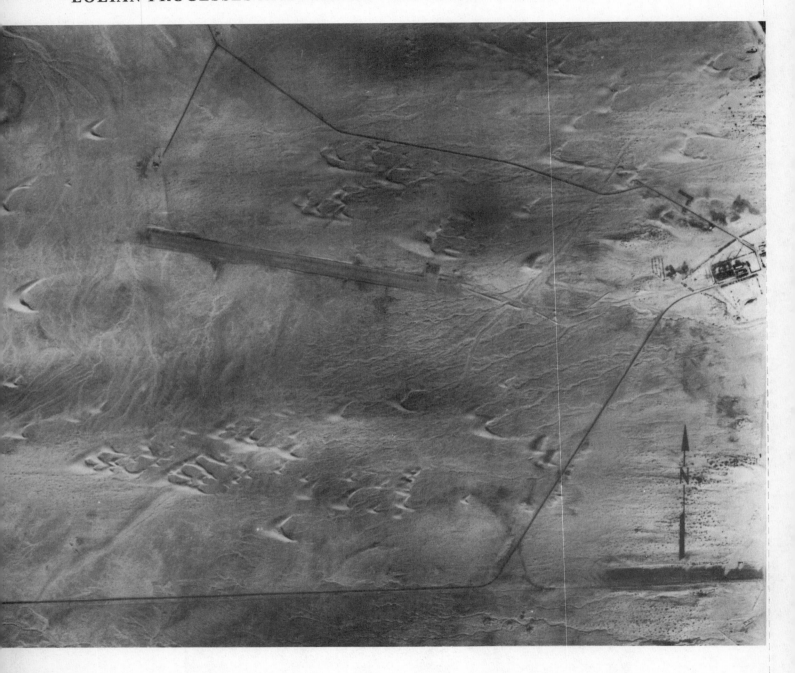

1. Name the type of dunes on the photos _____

2. What is the direction of prevailing winds? _____ How can you tell? _____

NAME _____

EOLIAN PROCESSES AND LANDFORMS – WASHINGTON

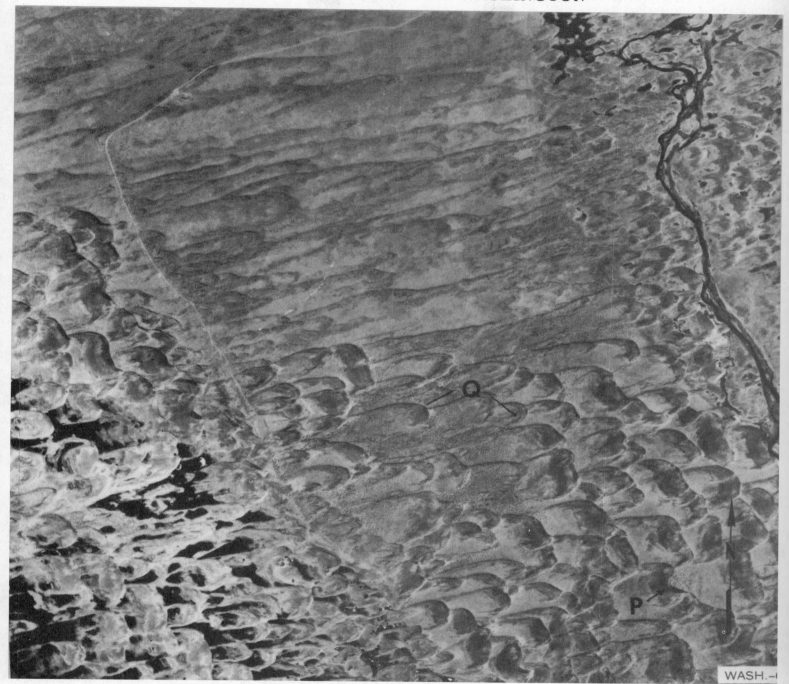

WASH.–

1. What is the direction of prevailing winds? (Look for slip faces) _____
2. Name the type of dunes at 'P'. _____
3. Name the type of dunes at 'Q'. _____
4. How can you tell the two types of dunes apart? _____
5. Why are they different? _____

EOLIAN PROCESSES AND LANDFORMS – ARIZONA

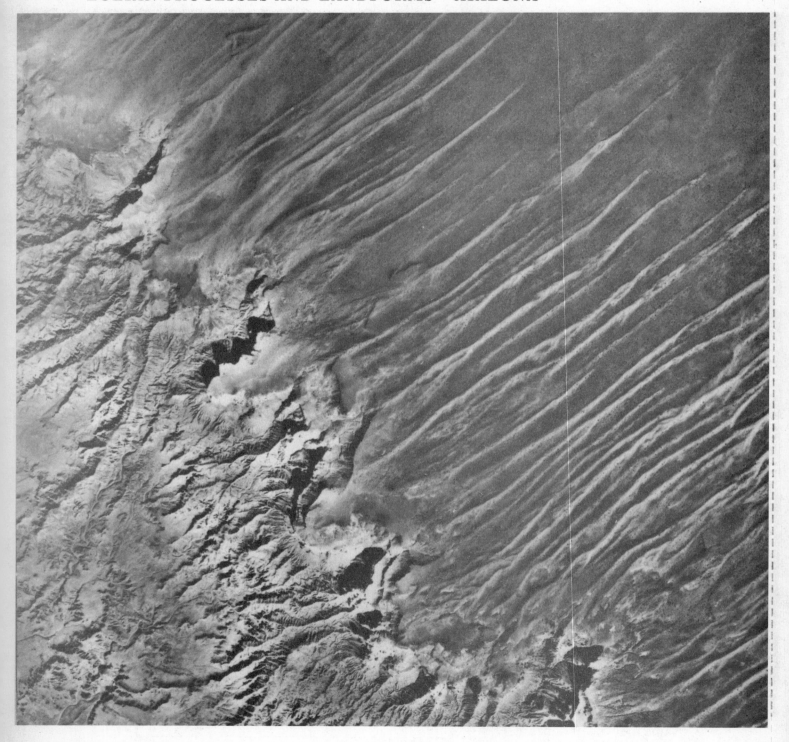

1. Name the type of dunes shown on the photos. _____

2. What controls their orientation? _____

EOLIAN PROCESSES AND LANDFORMS – PAKISTAN

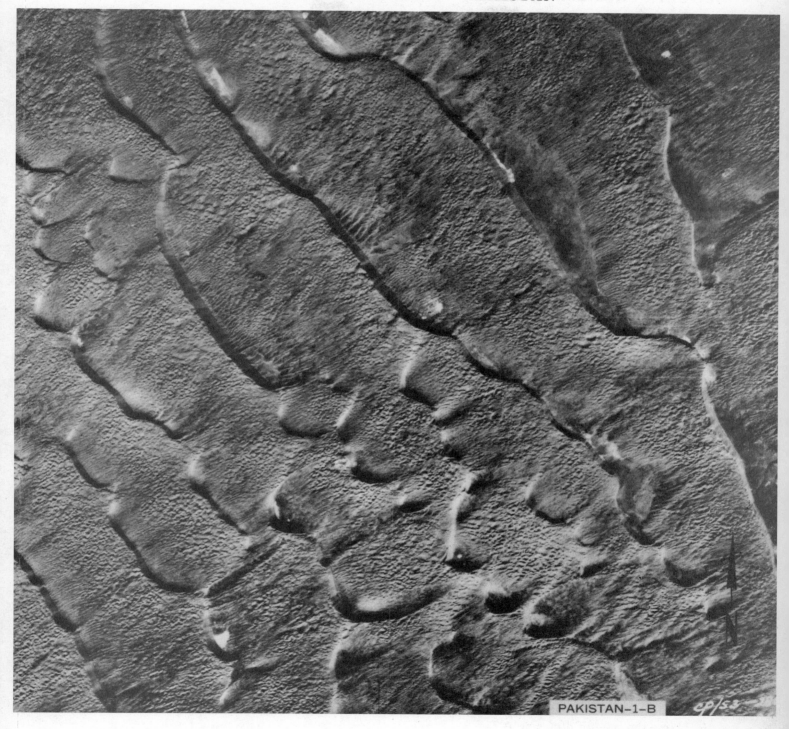

PAKISTAN–1–B

1. Name the dominant type of dunes on the photos. _____

2. What other type of dune occurs on photo? _____

3. What is the direction of prevailing wind? _____

EOLIAN PROCESSES AND LANDFORMS – ARIZONA

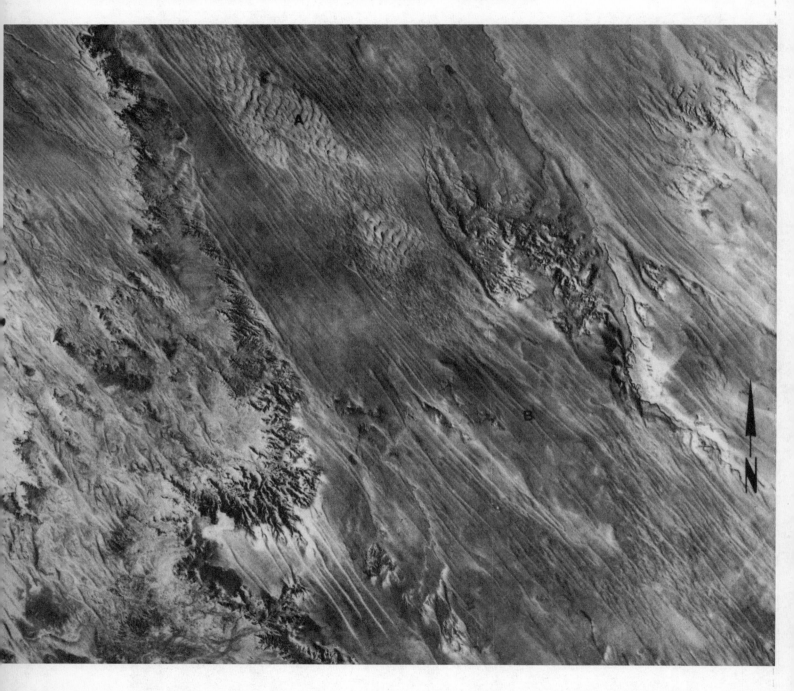

1. Name the type of dunes at 'A'? _____

2. Name the type of dunes at 'B'? _____

3.   What is the direction of prevailing wind? _____ How can you tell?
_____

Dating Geomorphic Features

Topographic Criteria for Dating Landforms

Relative Dating Methods

Cross-cutting landforms

Degree of stream dissection

Degree of weathering and soil development

Superposition of landforms

Numerical Dating Methods

Radiocarbon

K-Ar

Cosmogenic

Fission-track

Thermoluminescence

Amino acid

Uranium series

Paleomagnetism

Obsidian hydration

Tree rings, lichens

Radiocarbon-dated logs protruding from a volcanic mudflow, Nooksack River, Washington.

DATING GEOMORPHIC FEATURES

The diagram on the next page shows three moraines and several loess layers containing volcanic ash whose ages are shown. The ages of the three glaciations can be determined by analyzing the relationship of the moraines to the loess and ashes. Basal peat in the bog behind the youngest moraine is 14,000 years old.

1. What *relative dating techniques* could you use to establish the age difference between the moraines?

   a. _____   c. _____
   b. _____   d. _____

2. The older of the two volcanic ashes is 600,000 years old and the younger is 200,000 years old. What *methods* and what *materials* are *best* suited for dating such ashes?

Method	**Dateable Materials**	**Age range of the method**
a. _____	_____	_____
b. _____	_____	_____

3. What methods could be used to determine numerical ages from erratics on the moraines?

   a. _____
   b. _____

4. By what *method* might the loess be dated directly? _____

5. What *method* is best suited for dating the peat behind the youngest moraine? _____

6. Considering the distribution and ages of the loess and ash, what can you say about the age of

   the oldest moraine? _____ Evidence? _____

7. The age of the middle moraine must be older than _____ but

   younger than _____ Evidence? _____

   _____

8. The age of the youngest moraine must be older than _____

   but younger than _____ Evidence? _____

   _____

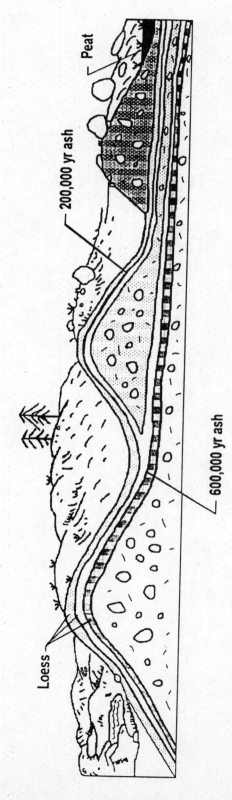

Appendix

List of Quadrangles

List of Quadrangles (Page 1 of 2)				
State	Quadrangle Name	Scale	Contour Interval (Feet)	Year Surveyed
ARIZ.	Gila Butte	1:62,500	20	1952
CA.	Jellico	1:62,500	40	1957
	Mt. Dome	1:62,500	40	1950
	Mt. Whitney	1:62,500	80	1956
	Tulelake	1:62,500	20	1951
	Medicine Lake/Timber Mtn	1:24,000	40	1988
	Prospect Peak	1:62,500	40	
CO.	Holy Cross	1:62,500	50	1949
	Loveland	1:24,000	20	1060
	Masonville	1:24,000	40	1971*
FL.	Crystal Lake	1:31,680	10	1945
	Jacksonville Beach	1:24,000	10	1948
	Lake Wales	1:24,000	5	1952
	Melbourne East	1:24,000	5	1949
ID.	Bellevue	1:62,500		1957
	Craters of the Moon Nat'l. Monument	1:31,680	40 and 20 (Dotted Lines)	1957
HI	Kilauea	1:62,500	50	1946
	Mauna Loa National Park	1:62,500	50	1968
KY.	Frankfort East and West	1:24,000	20	1950
	Little Muddy	1:24,000	20	1945
	Park City	1:24,000	10	1954
	Smiths Grove	1:24,000	10	1954
LO.	Caspiana	1:62,500	20	1955
MASS.	Chatham	1:24,000	10	1961
	Edgartown	1:24,000	10	1972
	Provincetown	1:24,000	10	1958
	Tisbury Great Pond	1:24,000	10	1951
	Westport	1:24,000	10	1963
MISS.	Greenwood	1:24,000	5	1982
	Schlater	1:62,500	5	1961
NM.	Ship Rock	1:62,500	20	1953
PA.	Milton	1:62,500	20	1953
	Williamsport	1:62,500	20	1986
R.I.	Kingston	1:31,680	10	1942

Note: * = Photorevised

List of Quadrangles
(Page 2 of 2)

State	Quadrangle Name	Scale	Contour Interval (Feet)	Year Surveyed
TEX.	La Paloma (1936)	1:31,680	1	1936
	La Paloma (1970)	1:24,000	5	1956, 1970*
	West Brownsville (1936)	1:24,000	1	1936
	West Brownsville (1970)	1:24,000	5	1956, 1970*
Utah	Hurricane	1:62,500	40	1954
VA.	Strasburg	1:24,000	40	1966, 1972*
WA.	Barnes Butte	1:24,000	10	1968
	Coulee City	1:24,000	10	1965
	Hamilton	1:62,500	100	1952
	Holden	1:62,500	100	1944
	Jameson Lake	1:24,000	10	1965
	Mt. Baker	1:24,000	40	1989
	Mt. Baker	1:62,500	100	1952
	Mt. Shuksan	1:24,000	40	1985
	Park Lake	1:24,000	10	1965
	Sims Corner	1:24,000	10	1968
WIS.	Sun Prairie	1:62,500	20	1905
	Waterloo	1:24,000	10	1959, 1971*
	Watertown	1:62,500	20	1959
	Whitewater	1:62,500	20	1960
WYO.	Antelope Ridge	1:24,000	20	1952, 1978*
	Burlington	1:24,000	20	1951
	Grand Teton Nat.'l park	1:62,500	50	1948
	Grass Creek Basin	1:62,500	25	1913
	Meeteetse	1:62,500	25	1911
	Otto	1:24,000	20	1951
	Spence	1:24,000	20	1966
	Sheep Mountain	1:24,000	20	1951
	YU Bench NE	1:24,000	20	1951
	YU Bench NW	1:24,000	20	1951
	Wolf Point	1:24,000	20	1954

Note: * = Photorevised